LOCAL BREW

*Traditional Breweries
and their Ales*

LOCAL BREW

*Traditional Breweries
and their Ales*

MIKE DUNN

ROBERT HALE · LONDON

© Mike Dunn 1986
First published in Great Britain 1986

Robert Hale Limited
Clerkenwell House
Clerkenwell Green
London EC1R 0HT

British Library Cataloguing in Publication Data
Dunn, Michael. *1948–*
Local brew: traditional breweries and
their beers.
1. Breweries – Great Britain
I. Title
663'.3'0941   TP573.G7
ISBN 0-7090-2793-1

Photoset in Palatino by
Rowland Phototypesetting Limited, Bury St Edmunds, Suffolk
Printed in Great Britain by St Edmundsbury Press Limited
Bury St Edmunds, Suffolk
Bound by WBC Bookbinders Limited

Contents

	List of Plates	7
	List of Line Illustrations	9
	Acknowledgements	13
1	A Local Tradition	17
2	Historical Development of the Brewing Industry	22
3	Breweries and Beers	38
4	South-East England	57
5	Southern and South-West England	80
6	Wales	108
7	The West Midlands	118
8	The East Midlands and East Anglia	141
9	North-West England	173
10	North-East England	204
11	Scotland	220
12	Offshore Breweries	228
	Bibliography	241
	Index	245

Plates

Between pages 64 and 65
1. Harvey's Brewery in Sussex
2. Ruddle's Brewery in Rutland in the early twentieth century
3. Theakston's Brewery in Caldewgate, Carlisle
4. Jennings Brewery and Cockermouth Castle in Cumbria
5. The cask yard and malthouse at Morland's
6. King & Barnes Brewery, Sussex, in 1900
7. The bottling hall at Davenport's, Birmingham, in the 1930s
8. The coppers at Davenport's Brewery, Birmingham
9. The mash tuns at the United Breweries, Abingdon, in the 1920s
10. The top of Davenport's conical fermenters
11. Burtonwood's cooper at work before his retirement in 1984

Between pages 128 and 129
12. Delivery by horse-drawn dray at Adnams' brewery tap, the Sole Bay Inn
13. Thwaites' shire horses on show
14. The Gipsies Tent, a former home-brew pub
15. Shakespearian quotations outside Batham's brewery tap, the Vine
16. Early transport at Ruddle's
17. Hook Norton – a classic tower brewery
18. The cover of an early twentieth century Hall & Woodhouse price list
19. Donnington Brewery in the Cotswolds
20. Simpkiss Brewery in Brierley Hill, West Midlands
21. Beard's brewhouse in Lewes, Sussex
22. The mash tuns at Davenport's in Birmingham
23. The mash tun at Harvey's in Lewes, Sussex
24. Fermenting vessels at Harvey's

25 Vigorously fermenting beer in Davenport's fermenting room
26 The miniature tower brewery at the Three Tuns in Bishops Castle, Shropshire
27 The changing face of brewing

Line Illustrations

	page
Hall & Woodhouse's 'new premises': the architect's drawing in 1899	Frontispiece
Hanson's brewhouse, Kimberley, Nottinghamshire in about 1899	18
Steward & Patteson's Norwich brewery	28
One of the many fine beers produced by the former Yorkshire Clubs' Brewery	35
A beer mat from Melbourns of Stamford	35
The brewing process	45
The former Union Room at the Bass Brewery, Burton-on-Trent	48
Beer engine fittings from the 1899 catalogue of T. Heath	52
Bottling at the Wrexham Brewery in the 1890s	56
Prize Old Ale – one of the few naturally conditioned bottled beers	63
A beer mat from King & Barnes of Sussex	67
McMullen's Brewery in Hartham Lane, completed in 1893	69
A McMullen's poster of 1931	71
A Young's advertisement	78
Donnington Brewery beer and stopper labels from the 1940s	87
A beer mat and label for two distinctive ales from Hall & Woodhouse	93
A Badger ales price list from the late nineteenth century	95
The Old Brewery, Bridport, Dorset	101
The auction notice for the Sun Inn at St Austell	103
The Old Brewery buildings of 1713	110
The East Yard at the Llanelli Brewery in 1890	113
A Felinfoel bitter label	117
The Banks's and Hanson's logo	121
A Batham's beer mat and 'Sparkling Pale Ale' label	124
A Holden's beer label	128
Marston's logo commemorating 150 years of independent brewing	132

	page
The Old Swan 'home-brew' logo	135
Beers of the past: the Simpkiss brewery was closed in 1985	137
An Adnams beer label	142
Renewed interest in traditional beers from Greene King: an advertisement from the late 1970s	150
A Hardy's showcard from the 1920s	153
The 1930 merger agreement between Hardy's and Hanson's	155
A Home Brewery beer label	156
A Hoskins of Leicester advertisement	159
Labels for Rayment's bottled beers – a vanished breed	164
An advertisement for Shipstone's bitter	168
A label from pre-1972 State Management Scheme days	174
Two Burtonwood beer labels	179
The mashing-stage at the Groves & Whitnall brewery in Regent Road, Salford in 1890	182
Higson's advertising	186
An architect's plan of the Derby Brewery in the 1930s	189
Jennings Brewery and Cockermouth Castle	192
An advertisement for Mitchell's beers	196
Oldham Brewery's offices, constructed around 1910	198
A Cameron's beer advertisement	205
A Taylor's beer label	212
A distinctive Theakston's beer mat	214
A label for Home Brewed Stout from the Puzzle Hall Brewery, taken over by Ward's in 1935	218
A Belhaven beer mat	222
A label for Mackay's well known 'Oatmalt Stout'	226
A beer label from the Guernsey Brewery	230
A beer mat advertising the wares of Randall's of Guernsey	232
A striking barrel label from Okell's Falcon Brewery	237
A Burt's Strong Brown Ale label	239
Map of Britain's local breweries	16

Acknowledgements

I am grateful to the following brewery companies for providing photographs: Adnams (12), Burtonwood (11), Davenport's (7), Hall & Woodhouse (18), King & Barnes (6), Morland (5, 9), Ruddle's (2, 16) and Thwaites (13). The remainder of the photographs are mine, though I am particularly indebted to Tony Muntzer, Brewing Director of Davenport's Brewery Ltd., and Miles Jenner, Head Brewer of Harvey & Son (Lewes) Ltd., for allowing me the freedom of their brewhouses for photographic purposes.

In addition I must record my thanks to those brewers who were kind enough to allow me to use line illustrations; special thanks are due in this respect to Boddingtons', for the line drawing of Oldham Brewery's offices from *The History of Oldham Brewery*; Brain's, for the illustration from *The Cardiff Brewery*; Eldridge Pope, for the diagram of the brewing process; Hall & Woodhouse, for the line drawing of their 'new premises' in 1899 and the nineteenth-century price list, from Hurford Janes' *Hall & Woodhouse 1777–1977*; Hardys & Hansons, for the illustrations taken from pages 22, 89 and 93 of George Bruce's *Kimberley Ale*; Higsons, for the drawing from *Higson's Brewery 1780–1980*; Joseph Holt, for the plan of the Derby brewery in the 1930s, from Christopher Grayling's *Manchester Ale and Porter*; Jennings, for the line drawing from *Jennings Country*; McMullens, for the poster and the line drawing of the brewery from *The McMullen Years*; Palmers, for the brewery illustration; St Austell, for the auction notice taken from Clifford Hocking's *St Austell Brewery*; and Wards, for the illustration of the Puzzle Hall brewery, from *Innwards*.

Mike Dunn

For Chris, Kate and Sarah

Britain's Local Breweries

1 A Local Tradition

One of the delights of travelling through the regions of Britain is to discover the distinctive products of local breweries, traditionally made and carefully attuned to local taste. This survival of local brewing enterprise is a very British phenomenon, too – only in Bavaria is anything like the same rich choice of beers from local breweries available. And the tradition has survived despite all the problems encountered in recent years – the unwelcome attentions of larger brewers, high prices and declining sales in pubs and so on. Some eighty long-established local brewers, including four historic home-brew pubs, are still in production, and they form the core of this book. Thankfully a high proportion of these local brewers remains committed to the use of traditional methods and ingredients to impart a local and distinctive flavour to their beers.

This reliance on traditional methods is hardly surprising, for brewing is a relatively unchanging process, and the monastic brewers of medieval times would probably feel thoroughly at home in most present-day local breweries. The monks were, in fact, critically important in developing the craft of brewing, laying the foundations in places as diverse as Burton-on-Trent, Oxford – where the Morrell family, in occupation of the Lion Brewery since 1782, are relative newcomers on a site where the monks of Osney were brewing in the fifteenth century – and Faversham, where Shepherd Neame are the successors of the Cluniac monks, who were producing beer there in the twelfth century.

Amongst the attractions, in many cases, are the architectural delights of the brewery buildings themselves. The best, in terms of their contributions to the townscape, include Wadworth's redbrick Victorian brewhouse, at the northern entrance to the market-place in Devizes, and Elgood's dignified late eighteenth-century brewery, in the Fenland town of Wisbech. Other outstanding Victorian edifices include Okell's ornate brewery in the Isle of Man, Harvey's in Lewes and Eldridge Pope's in Dorchester – this one later extensively damaged by fire but rebuilt in the original style.

Hanson's brewhouse, Kimberley, Nottinghamshire in about 1899

Inside the breweries time-honoured methods are still used to brew the beers – and very often traditional methods, in traditional vessels, are retained even in newer breweries. Brewing is a classic case of ancient methods proving the best, even to the extent where Hook Norton's Victorian steam-engine, for example, can prove as effective as modern successors. Yet within this general regard for tried and tested methods there have always been important local variations in practice: in and around Yorkshire, for example, square fermenting vessels made of slate (instead of the more usual stainless steel) are still used by Darley's, Samuel Smith and others, whilst further south Marston's are now the only brewery using the Burton Union system of fermentation. Equally unique are Lorimer & Clark's coal-fired coppers, whilst open cooling troughs are still used by a handful of brewers such as Elgood's and, amongst the surviving handful of home-brew pubs, the Old Swan (Ma Pardoe's) in the Black Country.

In the brewery cellars tradition is at work in a few worthy breweries where wooden casks are still the order of the day (though their effect lies less in improving the beer than in adding a picturesque touch to the brewery's operations). Wadworth's and Lees are amongst those who favour beer from the wood for most of their draught-beer production, whilst Samuel Smith's are so committed that they actually have coopers making new wooden casks as well as repairing old ones. Still more nostalgic appeal is injected by the use of horse-drawn drays to make local deliveries, and many brewers, including Adnam's, Vaux and Young's, now rely on horses to ferry beer to nearby pubs, and others maintain teams of shirehorses for special occasions. Nostalgia and publicity are not the only reasons for the return of horse-power, however, for it has been shown that horse-drawn drays are easily the most efficient and economic means of making deliveries of cask beers within a mile or two of the brewery.

The beer in those casks, whether they be wooden or metal, will very often have an aroma, a flavour and an aftertaste which are special to the brewery concerned – though it may share a regional characteristic with its local rivals – and which amply repays the effort of tracking it down and acquiring a taste for it. This may be due in part to the local water, which can have a decisive part to play in determining flavour (hence the rise of Burton-on-Trent as a brewing centre in the nineteenth century when it became clear that the water there was ideal for pale-ale brewing); it may be the yeast, which in many cases will have been in use in the brewery for decades. Whatever it is, the result is likely to be highly distinctive: hence the special taste of Adnams' beers, ascribed by many drinkers to the Suffolk seaweed! Or Batham's light-coloured, sweetish bitter, 'sugar water' to some but revered by many others. The loyalty of drinkers to their local brew is legendary: in many East Anglian towns the determination of Greene King drinkers not to cross the road and sample the Tolly Cobbold ales is matched only by the refusal of Tolly regulars to try Greene King's beers. Sometimes the loyalty is misplaced, as when Hardy's and Hanson's merged in the 1930s. For years regulars in Hardy's pubs would drink only Hardy's ale, whilst Hanson's drinkers insisted upon Hanson's – despite the fact that the two beers were now identical!

This local loyalty is also reflected by the brewers themselves, who are often keen to emphasize their ties with the area. This may take the form of naming beers in honour of local people – Wadworth's Farmer's Glory, for instance, recalls a novel by the Wiltshire writer A. G. Street, whilst Tinners Ale from St Austell commemorates the dying breed of Cornish tin-miners. Or it may be something more tangible, as with John Willie Lees' support for Oldham Athletic football club, the Vaux Group's links with the local television station, or Burtonwood Brewery's interest in Haydock Park racecourse. Other support for local institutions reflects the earlier involvement of many members of the 'beerage' in the corporate life of their local communities: the list of brewers who have achieved mayoral status is a long one, emphasizing the high social profile of the occupation.

But it is in the pubs and the beers which are drunk in them that the real glory of the local brewers lies. Some of the pubs may have few frills, some of the beers (including some of the very best) may shock the palate, but Britain would be a poorer place without the local brewers who provide colour in a drab world and immeasurably improve the range of choice and taste in our pubs. The story of the growth of our local brewers is never dull, including as it does the tale of the Reverend James Buckley, staunch disciple of Wesley yet founder of a brewing dynasty; of Charles Wells, sea captain, who was forced to buy a brewery in Bedford and settle down in order to win the approval of his prospective father-in-law; and of the late, lamented Doris Pardoe, for many years the landlady of one of Britain's finest pubs, yet a teetotaller who never touched a drop of the magnificent home-brew from the brewery behind the pub.

2 Historical Development of the Brewing Industry

The brewing industry today comprises seven national brewers (two of them merely subsidiaries of even bigger organizations); a handful of regional brewers such as Vaux and Greene King, operating two or more breweries; about seventy long-established local brewers ranging from tiny firms such as Batham's (with only eight pubs) to substantial local businesses such as Burtonwood Brewery and Eldridge Pope; about a hundred newly established concerns (i.e. 1972 or later) competing almost exclusively in the free trade (and with a very high turnover as new firms are formed and older firms go out of business); and a burgeoning number of home-brew pubs, of which only four can claim any real longevity. It is an unusual structure, brought about in large measure by historical accident and fossilized by restrictions which make it peculiarly resistant to real change.

There are three essential factors which do much to explain the present structure (some would describe it as the present sickness) of the industry. First and foremost is the growth of the larger firms – begun in the eighteenth century, when the rising popularity of a new product, porter, precipitated the rapid advance of the more imaginative and aggressive 'common brewers', and brought to its climax in the 1960s when 'merger mania' created, for the first time, truly national brewers with vast tied estates and, between them, a dominant market share. Secondly, the much earlier scramble to ensure sales by establishing a secure basis of tied houses, which was caused by fairly draconian licensing restrictions and was at its height in the 1890s, created a property-owning 'beerage' with the power to reduce or exclude competition. Thirdly, the exercise of that power (and its concentration in fewer hands as successive bouts of take-over fever took their toll), in conjunction with restrictive licensing and planning legislation, has contributed to a situation where fewer and fewer beers are produced, for a stock of public houses which is either formally tied to the products of one brewer or, in the case of 'free' houses, very often 'loan-tied' to the bigger brewers, who have used their financial muscle to ensure that their (usually inferior) products are stocked to the exclusion of the generally excellent beers of their local rivals.

In the beginning

Such considerations were far from the mind of Britain's earliest ale-brewers, who inherited their knowledge of the necessary arts of cereal cultivation, malting and fermentation from their Middle Eastern forbears. Archaeological evidence suggests that barley was cultivated in Mesopotamia by 5000 BC and that beer brewed from malted barley had become part of the staple diet there about 2,000 years later. Comparative evidence for Britain indicates that cereals were being cultivated by about 3000 BC, but the earliest evidence of brewing is much later: certainly the practice was firmly established by the time of the Roman invasion.

Fermented beverages were crucially important in Dark Age and medieval Britain, given that water was commonly unfit to drink. By the tenth century, indeed, beer was so popular that a decree was issued limiting the number of alehouses to one per village. Four hundred years later three types of establishment catered for drinkers: inns, which also offered food and accommodation; taverns, which also sold wine; and, most basic of all, alehouses. Between them these premises accounted for only a tenth of consumption, however, for most ale was brewed and consumed in the home, especially in the more rural areas of the country.

Until about 1400 all beer was actually 'ale', an unhopped beverage of which water, malt and yeast were the only permitted constituents. But in the early fifteenth century hops were introduced from Flanders, primarily as a preservative (though they also had a very considerable effect on the flavour and aroma of the product, and it was also possible to produce a weaker brew, since the only way to make ale keep had been to brew it with a very high alcohol content). There was considerable resistance to the addition of hops to the brew – as late as 1519 brewers in Shrewsbury were prohibited from using the new-fangled ingredient – but the acceptance of hopped beer, which was weaker, more bitter and generally of better quality than ale, gradually spread throughout the country. By the eighteenth century 'ale', originally spiced and leavened with honey, and therefore sweet, thick and cloying, had finally given away to the 'beer' whose descendants still grace the bars of Britain today.

The rise of the common brewer
In the later medieval period most commercial ale- and beer-production was in the hands of publican-brewers, but with the growing popularity of hopped beer the first 'common brewers' (wholesale brewers who supplied their beer in bulk to publicans instead of selling it all retail) came into existence. Whilst their main trade continued to be in unhopped ale, their expansion was strictly limited, since their product had to be sold quickly before it began to deteriorate, and therefore was restricted to the immediate, local area of the brewery, but once hopped beer became the mainstay of their business, the more adventurous common brewers seized the opportunity to expand production and compete in a wider market. Publican-brewers held their own for some time, so that even in London most beer was still brewed on the premises in 1600, but by the end of the seventeenth century there were 194 common brewers in London, and their combined production of 1,650,000 barrels a year dwarfed the 12,000 produced by the remaining publican-brewers.

Changing public tastes also contributed to the rise of the common brewer, none more so than the sudden growth in demand for porter in the 1720s. The new beer evolved from 'three threads', a drink which had to be mixed from separate casks of new and stale brown ale and pale, much to the irritation of publicans. In 1722 Ralph Harwood, of the Bell Brewhouse in Shoreditch, developed a beer reputed to combine the merits of all three constituents of 'three threads', and called it 'entire' since it was drawn entirely from one cask. The new dark, thick and heavily hopped beer quickly developed considerable popularity, especially amongst London labourers and porters – hence the change of name. The beer was more expensive to produce, since it had to be matured over a longer period (up to a year in oak vats in some breweries), and it was therefore impractical for publican-brewers to become involved: the field was open for common brewers to dominate this new trade and – through deriving economies of scale in the long run – to attain a higher profit.

Porter-brewing quickly spread throughout the country, to Sheffield by 1744, to Bristol before 1788 (when the Bristol Porter Brewery was established by Philip George) and to Glasgow by

1795. It was described as 'the universal cordial of the populace' in 1758 – and this despite the fact that the best malt and hops were rarely used, and its clarity was of little importance, since its almost black colour hid most imperfections. Yet by the 1830s demand was declining, and most leading porter brewers, such as Whitbread, were turning to pale-ale brewing by this time. Forty years later most of the leading porter-brewers in London had recognized the change in demand by setting up pale-ale breweries in Burton-on-Trent, where the water was regarded as ideal for lighter beers. Ind Coope established a brewery there in 1856, Charrington's in 1871 and Mann, Crossman & Paulin built the Albion Brewery (now the home of Marston's) in 1875.

The heyday of porter coincided with the establishment of the first volume breweries, rising out of the ranks of the common brewers. Few London brewers were producing more than 5,000 barrels a year in 1700, but by 1800 three of them – Barclay Perkins, Whitbread and Meux Reid – were brewing 200,000 barrels or more. Whitbread had led the way to 200,000 barrels, reaching that target in 1796, but they were soon overtaken by Barclay Perkins, who were the first to 300,000 barrels (in 1815). The spectacular speed of growth was inevitably accompanied by concentration: as early as 1748 half of London's beer was produced in just twelve brewhouses.

The eighteenth century saw the birth of many of the concerns which were later to grow into the major forces in the brewing industry. Amongst the London brewers were the Stag Brewery in Pimlico, established as early as 1636, rescued from financial disaster by Elliott in 1787 and eventually taken over by Watney's; Whitbread's, established in Old Street in 1742 and growing so rapidly that a porter brewery was built in Chiswell Street eight years later; Charrington, brewing at the Anchor Brewery in Mile End from 1766; and Courage, established at Horselydown in Southwark by 1789. At Burton-on-Trent, Samuel Allsopp's was founded in 1709, Worthington in 1744 and Bass in 1777.

Between 1750 and 1800, the number of common brewers rose by forty per cent, to 1,382, indicating the extent to which the market was expanding (and, conversely, the speed of decline of publican-brewing). Later, however, growth was less rapid, for new entrants to the industry were immediately at a commerical disadvantage, unable to achieve the economies of scale of their established competitors – who also acted at times to destroy unwanted competition, most frequently by making every effort to exclude them from the tied trade, which, as we shall see, was becoming more prevalent as the nineteenth century progressed.

The growth of the tied house system
The early prosperity of the common brewers owed much to their wholesale status. It was unusual for them to own tied houses since they had neither need nor desire for them, regarding them as unwanted encumbrances which tied up part of their working capital without providing a tangible return. This philosophy endured whilst there was a predominantly free trade (though one in which many publicans habitually obtained their supplies year after year from the same brewer) and a licensing system which permitted new alehouses to be opened at will. Given these conditions, common brewers made no effort to build up a tied estate, taking over alehouses only through the default of publicans to whom they had loaned money on mortgage (and even as early as 1686 many publicans are known to have been in debt to their suppliers).

The closing decade of the eighteenth century saw a combination of factors which altered the brewers' outlook, notably a toughening of the attitudes of the Justices of the Peace, poor harvests, war taxation and an unprecedented decline in demand. A very rapid increase in the number of tied houses ensued, especially in London, so that by 1810 Truman's supplied 481 pubs, of which no fewer than 378 were tied; Thrale's 477, with 277 tied; and Whitbread 308, with 256 tied. The House of Commons Committee on Public Breweries, reporting in 1819, noted this trend with obvious dissatisfaction, commenting that they 'objected to the monopoly which the brewers were creating'.

The increasing scale of the largest brewers, and the vogue for setting up breweries in Burton-on-Trent (always an important centre, but with its status enhanced even more with the coming of the railways and the spectacular growth of brewers such as Worthington and Bass) led to the flotation of many as public companies, increasing their financial scope still further and pointing the way to national prominence in some cases. The value of 'free' alehouses doubled or even trebled as they were fought over by the bigger breweries: in 1888 it was noted that, 'There had been a tremendous run on free houses, and in the course of a short time there would be scarcely a house worth buying in the market'.

Almost everywhere the publican-brewer was in retreat, and his virtual extinction was near at hand.

The growth of local breweries

The late nineteenth century saw dramatic growth not only in the largest common brewers but also in concerns which were then much smaller and were also destined to remain local in outlook – never, perhaps, extending much beyond the traditional thirty-mile trading area dictated by the day's range of a horse-drawn dray. Such local breweries were forced to compete for tied houses, however, in order to secure their niche in the limited area they had made their own. Brain's, for example, supplied only eleven public houses in 1882, but fifteen years later they controlled seventy-four (all very close to the brewery, a situation which with only a handful of exceptions still pertains today) and had increased their output tenfold. Joseph Holt's were equally drawn in to the mad rush to buy tied houses at the end of the nineteenth century, and indeed they more than doubled their tied estate with the purchase of no fewer than forty-two pubs in the four years from 1896.

At the same time many of the brewing families, presiding over this extraordinary expansion and acquiring both wealth and prestige, became heavily involved in public service, sometimes at the risk of neglecting their business interests. The Holt family were particularly active in Manchester life, with Edward (later Sir Edward) Holt a magistrate and a member of the city council for over thirty years in addition to his work for the Manchester Brewers' Association, of which he was vice-president. He was, unusually, Lord Mayor of Manchester for two consecutive years (1907–9) and was a driving force behind the scheme to bring water to Manchester by pipeline from Haweswater Reservoir in the Lake District.

A similar story can be told of the Brakspear family, who first assumed sole control of the Henley Brewery in 1803. William Henry Brakspear, son of the founder, was a justice of the peace, alderman and four times mayor of Henley-on-Thames; in their time his father and son both served as mayor, and the founding Brakspear had also preceded him as JP and alderman in addition to achieving still further status as a churchwarden of the parish church. Though Brakspear's still faced the wrath of temperance organizations (which were sometimes remarkably effective, even to the extent that a newly built Brakspear pub had to survive from its opening in 1899 to 1930 as a house licensed to sell mineral waters only), the occupation of brewer was clearly one of some prestige in Victorian times, not least because of the continual increase in trade as more and more publican-brewers bowed out of the business and their houses were snapped up by the common brewers, including the local firms which were slowly building up a sufficient tied-house base to sow the seeds of their survival in the increasingly competitive world of the twentieth-century brewing industry.

The demise of the publican-brewer
Publican-brewing, as was noted above, accounted for most beer production until the seventeenth century, even in London. Elsewhere common brewers were established later and grew more

Steward & Patteson's Norwich brewery – absorbed into the Watney group in 1963

slowly, so that even in 1800 the 40,000 surviving publican-brewers, together with private brewing, accounted for about half of total beer production. During the nineteenth century, however, the twin obstacles of taxation and competition drove well over three-quarters of the publican-brewers from the industry, and not long after 1900 the species had, sadly, become virtually extinct except in a small number of isolated areas such as the Black Country.

There were, of course, some exceptions to the rule of the common brewer in the nineteenth century: Leeds, for example, fostered the splendid tradition of home-brewed ale for longer than most cities, with the last survivor, the brewery behind the Nelson Hotel in Armley Road, possibly in operation up to the death of the owner, Samuel Ledgard, in 1952. But in general the decline of the publican-brewer was depressingly rapid. His eclipse has been so total (until the recent reversal of the trend with the move back to home-brewed real ale in many free houses) that in many cities there is not even a folk memory of publican-brewing. In Birmingham, however, with its history of individuality, independence and small-scale enterprise, the home-brew tradition carried on almost undisturbed for a remarkably long time: even in the 1880s the common brewers contributed only a quarter of total production. Indeed, ninety-seven per cent of Birmingham publicans held a brewing licence in 1873. Sadly the proportion dropped to less than one-fifth in the next twenty years as the common brewers moved unchecked into the extensive Birmingham beer market, and the tradition of publican-brewing collapsed with astonishing rapidity. Only the Dog and Duck Inn in High Street, Aston, which ceased brewing in 1921, seems to have survived well into the twentieth century.

By the end of the Second World War pubs which brewed their own beer were already extremely rare, and the succeeding years have taken a heavy toll of those which remained. In 1965 there were only four survivors, two of them in Shropshire (the All Nations in Madeley and the Three Tuns in Bishops Castle) and one each in the Black Country (the Old Swan in Netherton, near Dudley) and Cornwall (Blue Anchor, Helston). All four of these have continued in production to the present day, and in the last decade they have been joined by some eighty newcomers, welcome additions to the choice confronting the beer-drinker, but short on pedigree and many facing an uncertain future.

The most recent losses from the short list of long-established home-brew pubs have generally owed their demise to the death or retirement of the publican-brewer. This was the case with the Golden Lion in Southwick, Hampshire, where the proprietor, Mr W. J. Hunt, closed down the brewhouse in 1956 when he decided that he had become too old to keep it going single-handed; the pub (now a Courage house) survives, as do the brewery buildings, though these are now reported to be in a dangerous condition.

More recently the Gipsie's Tent in Steppingstone Street, Dudley, with a history of home-brew dating back to 1850 and latterly run by the Millard brothers, ceased brewing in 1966, becoming a free house and, later still, a forlorn and empty shell in the shadow of Hanson's brewery. Two years later the death of the licensee spelled the end of brewing at the Bluebell in Hockley Heath, Warwickshire, which became a Bulmer's cider house. In 1971 the Druid's Head in Coseley, near Wolverhampton, lost its home-brew when the brewer, Jack Flavell, died. The most recent, and in some ways the most poignant blow, since it could conceivably have been avoided, was the loss in 1972 of the excellent home-brew produced at the Britannia Inn in Pinfold Gate, Loughborough. Again the death of the licensee and brewer, Herbert North, brought about the end of the home-brew, in this case best bitter and mild, which had been produced since the days of Herbert North's father in 1923.

Decades of destruction: merger mania and the rise of the Big Six
The phenomenal growth of the most successful common brewers, and their rapid acquisition of tied houses, altered the face of the British brewing industry but nevertheless could be accommodated without too crippling a loss of choice because of the exceptionally competitive nature of the trade hitherto. Despite the emergence of major brewery companies, there were still 6,500 brewers (including publican-brewers and other tiny concerns) in Britain in 1900, and even in 1945 there were still 700, despite the effective end of publican-brewing. Fifteen years later, 358 breweries operated by 247 different companies were still in production: there was a high degree of local choice and local competition, and, although a handful of beers such as Worthington and Guinness had a national reputation, there were no truly national brewers in existence.

The 1960s changed all that, with the creation of a handful of national companies owning large numbers of breweries and public houses, and wielding (and using) unprecedented marketing power. Of the 247 independent companies which survived to 1960, some 150 failed to see out the decade. The majority were cynically taken over and closed down by companies interested only in extending their trading area and increasing their number of tied outlets. Some companies came together in defensive alliances to try to protect their independence, usually to no avail – as with Threlfall's and Chester's in the North-West, who merged in 1961 only to be snapped up by Whitbread six years later, or Cheltenham & Hereford Breweries (630 pubs) and Stroud Brewery (640 pubs), whose merger in 1958 created the mammoth West Country Breweries Limited, also victims of Whitbread, this time in 1963. Only a handful of breweries, other than family companies which controlled their own destiny, escaped the attentions of the predators, and only two who opposed formal offers emerged intact – Boddington's, who were the target of Allied Breweries in 1969, and Jennings, courted somewhat later by Mount Charlotte Investments.

147 separate breweries were closed in the 1960s: substantial breweries too, such as Hunt Edmunds of Banbury, bought up by Bass, Mitchells & Butlers in 1965 and closed down two years later, as well as 'smaller and older breweries' like those axed by Charrington in a single 'year of consolidation', 1964: Alloa, Brighton, Hull, South London (Woodhead's) and Yeovil. In the same period the market share of the seven largest brewers (Allied, Bass, Courage, Guinness, Scottish & Newcastle, Watney and Whitbread) had increased from forty-five per cent to seventy-five (it reached eighty per cent in 1972 and has since grown more slowly). Six of the seven – all except Guinness – had also built up, to a greater or lesser extent, very large tied estates, destroying local choice and substituting bland, insipid mass-produced beers which had to be supported by marketing efforts on an unprecedented scale.

The Big Six brewers were jolted into defensive amalgamations and substantial growth by the attempt of an outsider, Sir Charles Clore, to gain entry to the beerage by bidding for Watney's in 1959. In 1961 Ind Coope, Ansell's and Tetley joined forces to form Allied Breweries, the first truly national group; six years later the merger of Charrington United Breweries and Bass, Mitchell's &

Butler's created Bass Charrington, the largest brewery group in Europe. Meanwhile Courage (later bought by Imperial Group), Scottish & Newcastle, Watney (who fought off the Clore bid but were unable to thwart a take-over bid from Grand Metropolitan Hotels in 1972) and Whitbread all chose to indulge in a spate of take-overs to secure their place in the industry's elite band of national companies. Today Bass lead the way, with a market share of twenty-one per cent; Allied have fourteen per cent; Whitbread and Watney about thirteen apiece; Scottish & Newcastle ten per cent and Courage nine.

Real ale and the local revival
Whilst the 1960s saw a rapid contraction of the 'independent' sector of the brewing industry, the 1970s saw a virtual end to take-overs – certainly those involving the Big Six – and a resurgence of interest in the residual local brewers. There were a number of reasons for this astonishing reversal, the first of them political, with clear indications from Whitehall that any further monopolistic steps by the biggest brewers would be viewed with disfavour. Secondly, the economies of scale which were expected to result from the concentration of production in fewer and more modern breweries obstinately refused to materialize. Thirdly, the spectre of consumerism reared up before the brewers in the form of the Campaign for Real Ale (CAMRA), now acclaimed as phenomenally successful in turning back the keg tide and promoting not just real ale but also the beers of local brewers rather than the bland confections of the Big Six.

Real ale has proved to be a very useful bandwagon for many of the smaller independent brewers to hitch their wagons to, even though in many cases their notion of 'traditional draught beer' differs from that of CAMRA. Many of them rejected keg beer in the 1960s, but whilst this was for reasons of quality and tradition in some cases – as with Young's, who resolutely backed their distinctive cask-conditioned beers and reaped their reward as customers actively sought out their pubs – it was generally a combination of inertia and lack of finance which saw traditional beer retained. King & Barnes, for example, now admit that the cost of kegging equipment was beyond their means in the 1960s, leaving them ideally placed when the pendulum swung back towards real draught beer a decade later.

The new mood of determination to survive was indicated by

Morland's chairman in 1968: he saw a future for 'smaller breweries operating efficient outlets in limited areas'. Whilst this has been the general pattern, some of the most celebrated survivors have chosen very different ways of existence. Ruddles, for instance, have become one of the biggest suppliers of bottled and canned beers to supermarkets, a cut-throat and low-margin business but one in which they have a virtual monopoly of quality products. The cost of following this path (and ridding themselves of a substantial Whitbread shareholding) has been the loss of their thirty-eight pubs, sold in 1978 (not a popular move with the locals), but profits are rising and draught Ruddle's County is very widely available. Theakston's rise to prominence was also associated with a very distinctive real ale, Old Peculier, and they too ran into problems as they expanded distribution nationwide; cash shortages led to pub sales, outside finance, and eventually takeover by Matthew Brown.

Other regional brewers have prospered by following more conventional routes: Young's have already been mentioned, but their London neighbours Fuller's have been equally successful in expanding their tied estate and promoting excellent beers such as London Pride and ESB. Elsewhere, new traditional beers helped Eldridge Pope – who based their Royal Oak strong ale on an 1896 recipe and made it available only in traditional form – and Everard's, whose return to the real-ale fold was spearheaded by Old Original, whilst brewers such as Wadworth's, Shepherd Neame and Marston's had only to promote their excellent existing beers to benefit. And the national brewers were forced to join in, re-inventing the traditional beers which they had hoped to phase out altogether and even re-creating the names of companies they had taken over and closed down, in order to appear to have a more genuinely 'local' presence.

Recent times
The 1980s, however, have seen the return of the predators – not this time the national companies but the more ambitious local companies who have been busy buying out their rivals. It is not particularly difficult to find explanations for this surge in takeover activity. The continuing recession and associated decline in demand for beer (down from 40.6 million bulk barrels in 1978 to 36.7 million in 1984), the resultant over-capacity in the industry which meant that some brewers were producing far below their

optimum output (Border Breweries, for example, were brewing on only five days a fortnight just before they were taken over), and the ambition of some of the larger companies all contributed to an unstable situation.

The spate of 1980s take-overs has certainly made a modest contribution to the reduction of capacity: Border, Hull Brewery, Simpkiss and Yates & Jackson have all brewed their last pint. Several other firms soldier on as subsidiaries of their new owners, with their future in some doubt. Higson's, Theakston's and Davenport's come into this category, whilst Oldham Brewery, bought by Boddingtons' in 1982, seems particularly vulnerable now that their masters have increased their capacity still further with the purchase of Higson's and the Liverpool firm's new lager brewery.

So advanced is the process of concentration in the brewing industry that there are probably few, if any, surviving brewers who have not received a take-over bid. Hook Norton were receiving approaches as far back as the 1920s, whilst Timothy Taylor's received them through the post and adopted the admirable practice of filing them unanswered in the waste-paper basket. Northern Foods were widely believed to have approached most independent brewers in the 1970s, and in addition to their purchase of Hull Brewery (later sold to Mansfield Brewery) they certainly tabled bids for Oldham Brewery, Shipstone's (rejected by ninety-six per cent of the shareholders) and Tolly Cobbold, in whom they built up a twelve per cent stake only to see Ellerman Lines snatch control. There are three survivors, however, which spring to mind in showing that unwanted bids *can* be beaten off: Boddingtons', Jennings and Matthew Brown.

Boddingtons' thwarted Allied Breweries' north-western ambitions in 1970, a time at which local breweries assailed by take-over bids were expected to curl up and die. Their famous draught bitter was saved with the help of Colonel Whitbread, who is reputed to have said 'You are a very old firm. You have a very good name. You mustn't go out.' Jennings survived a bid from an erstwhile subsidiary of Grand Metropolitan, which snapped up the neighbouring Workington Brewery at the same time and would certainly have shut down the Cockermouth brewhouse. And Matthew Brown's sensational escape from the clutches of Big Six predator Scottish & Newcastle made the headlines in 1935.

One of the many fine beers produced by the former Yorkshire Clubs' Brewery

A beer mat from Melbourns of Stamford, who ceased brewing in 1974

Excluding the four long-established home-brew pubs, there are seventy-nine brewing concerns listed in later chapters. By no means all of them are entirely independent, however: a number are wholly owned subsidiaries which through some quirk of fate have been allowed to retain a separate identity, others are owned by firms from outside the brewing industry, and still more are partly owned by Big Six brewers (of whom Whitbread, with their own investment company, are the most notorious) or even by their own larger brethren. In fact there are only sixty-four separate firms.

Big Six shareholdings in local brewers are nowadays largely concentrated in the hands of Whitbread, whose spread of interests is awesome, though Bass also have a number of share stakes and there are other trading agreements. The Whitbread 'umbrella', originally promoted as a defensive policy of association with the aim of preserving the independence of local brewers, operated from the early 1950s, and by 1961 Whitbread had a minority shareholding in seventeen firms which between them owned 10,000 pubs (many of them, of course, now owned by Whitbread). Five years earlier the Whitbread Investment Company had been set up purely to manage these holdings. Now Whitbread or its investment company has major shareholdings (over twenty per cent) in Morland, Marston's, Brakspear's, Boddingtons' and Devenish, and substantial stakes (around ten to twenty per cent) in Buckley's, Fuller's, Matthew Brown and Hardy's & Hanson's, with smaller stakes in a wide range of other brewery companies. Bass has holdings of around thirty per cent in Castletown Brewery (inherited when Hope & Anchor Breweries was taken over by Bass) and Maclay's, one of only two Scottish breweries retaining local control.

Of sixty-four independent concerns, then, there is a substantial Big Six interest in twelve and a smaller stake in many more, yet there are still more inter-connections. Felinfoel, for instance, are very nearly a subsidiary of their Llanelli neighbours Buckley's, since Buckley's bought out one of the two controlling families in 1965, gaining 49.5 per cent of the shares, only to be rebuffed by the other family. Samuel Smith's have a considerable investment in Jennings, and an even odder link is the fact that ninety-two per cent of Hook Norton's ordinary shares are in the hands of directors of Burtonwood Brewery – though here there is no formal bond. And pairs of breweries such as Gale's and Fuller's, and Eldridge Pope and Palmer's, have directors in common.

On the other hand, there are still quite a number of totally independent brewery concerns, often family-controlled and sometimes with a mere handful of shareholders. Given commitment by successive generations, a reasonably progressive approach to marketing, and a distinctive range of beers, these local breweries can survive into the 1990s and beyond as living proof of the greatness of Britain's brewing heritage.

3 Breweries and Beers

Before looking individually and in some detail at Britain's local brewers, it is well worth looking first at the breweries themselves, to appreciate their visual impact in the landscape, and then inside the buildings, at the way in which beer is brewed and the different kinds of beer which are available.

Brewing in the landscape
The sight, the sound and (perhaps above all) the smell of a working tower brewery are unforgettable. The tall, narrow brewhouse, designed to make maximum use of gravity as the beer descends from stage to stage in the brewing process; the gentle hiss of steam, the rumble of wooden or metal casks in the brewery cellars, and (in some cases) the clip-clop of hooves as the horse-drawn dray begins its journey to local pubs; the pungent aroma of malted barley boiling with the hops (try walking through Cardiff's Royal Arcade when the nearby Brain's Brewery is in full flow!): these are the essence of British brewing.

But the impact of brewing in the landscape begins well before the ingredients even reach the brewery. The hop yards or hop gardens of Kent, Hereford & Worcester and Hampshire, often shielded by tall windbreak trees but instantly recognizable with their high poles and criss-crossing network of wires to prevent the crop trailing along the ground, have been an integral part of the landscape since the seventeenth century. Just as familiar in these parts of the country are the oast-houses, tall and attractive kilns where the hops are dried and compressed into six-foot hessian 'pockets'. Other areas are renowned more for their maltings, where barley is converted into fermentable malt; those at Donnington and Belhaven are disused now, but several breweries still run working maltings, such as those of Wolverhampton & Dudley Breweries at Lichfield and Langley.

Many of the surviving local breweries are worth a detour not just for the excellence of their beer but also for their contribution to the landscape. This applies not only to rural delights such as

Donnington Brewery, where the grouping of millpond, waterwheel, brewery and former maltings in a delightful Cotswold valley is quite stunning, but to such substantial urban breweries as Wadworth's, close to the market-place in Devizes, and Hardy's and Hanson's in Kimberley, and indeed to huge, imposing edifices such as that of Marston's in Burton-on-Trent, where the brewery has won awards for its sympathetic conservation of the Victorian buildings.

In some cases it is the distinctive architecture of the brewery buildings which is paramount. In Wisbech, for example, the noble Georgian façade of Elgood's brewery, dating from about 1790 and converted from a granary, is one of the outstanding features of the North Brink, that marvellous piece of townscape formed by the juxtaposition of domestic and commercial buildings lining the River Nene. But the zenith of brewery-building was the last quarter of the nineteenth century. Among the many fine late-Victorian brewhouses are Higson's in Liverpool (originally Cain's Mersey Brewery, conceived on the grand scale and always too big for the job in hand) and Okell's on the Isle of Man – a particularly ornate example dating from 1875. And the brewery building boom attracted architects of repute such as William Bradford, whose work can be seen deep in the Oxfordshire countryside, at Hook Norton, and in Lewes, where he built Harvey's new brewhouse in a strikingly similar Victorian Gothic style in 1881.

There are, too, breweries where the sense of history is as strong an attraction as the brewery itself. To take one example, the site of Morrell's Lion Brewery in Oxford has been used for making beer since at least the fifteenth century, when the monks of Osney Abbey were the brewers; during recent work to extend the brewhouse some of the medieval foundations have been uncovered near the Wareham Stream. Monks were the first brewers in other places too: at Faversham, Burton-on-Trent and Belhaven, where their two-foot-thick stone walls are thought to date from the fourteenth century. At Fuller's the Brewery House, down by the River Thames in Chiswick, occupies the site where beer was first brewed in the seventeenth century.

Whilst a few breweries contrive to look like ordinary, drab industrial buildings, these are heavily outnumbered by the picturesque and the unusual – by breweries, in fact, which exude character and hint that something special happens within their walls. Amongst the first group are Donnington Brewery, already mentioned as the proud possessor of the most beautiful brewery site in the land, its mellow stone buildings tucked into a quiet Cotswold valley, and Hook Norton, its striking Victorian tower brewhouse marvellously seen from across the fields to the south of the village. Just as dramatic is the situation of Guernsey Brewery, on the seafront in St Peter Port and at its best when viewed from Castle Cornet, on the other side of the island capital's bustling and colourful harbour. Much farther north, the brewery buildings of Jennings of Cockermouth lie immediately above the confluence of the rivers Derwent and Cocker, and immediately below the crumbling walls of Cockermouth Castle.

Palmer's of Bridport have the unique distinction of operating from a partly thatched brewery, and they are one of several with working waterwheels (Morrell's and Donnington are amongst the others). The identity of Bateman's of Lincolnshire was for decades bound up with the ivy-covered windmill next to the brewery, which featured on the whole range of advertising material. And the symbol most Adnam's enthusiasts would associate with their brewery is the Sole Bay lighthouse, featured on many beer labels and standing guard over the brewhouse in the seaside town of Southwold. Finally Batham's, with the idiosyncracy of true Black Countrymen, rely on Shakespeare, quoting his 'Blessing of your heart; you brew good ale' across the front of the brewery and the adjacent pub. Few, having sampled the beer in that pub, the world-famous Bull and Bladder, would argue with those sentiments.

Sadly, breweries such as this are increasingly rare nowadays. Gone are the days when Burton-on-Trent was dominated by the serried ranks of brewhouse towers, and when other important brewing centres such as Edinburgh boasted a large selection of brewing delights. The story of the demise of Maidstone's breweries is one which is typical of many British towns. Five of them survived to the 1950s; none of them stands now. Mason's Waterside Brewery, taken over in the 1950s, and the Metropolitan Clubs' Brewery, which became bankrupt at the same time, were soon demolished, but it was not until the 1970s that Isherwood's tall and imposing brewery in Lower Stone Street was torn down, to make way for a shopping centre. The Medway Brewery, once run by Style & Winch but latterly by Courage, has for some years been represented by a derelict site on the riverside, whereas the site of Fremlin's famous Pale Ale Brewery is now occupied by Whitbread's warehouse.

A more tangible though somewhat eerie link with the past is provided by the derelict (or sometimes re-used) breweries which still stand. The ghostly buildings of Campbell, Hope & King's fine brewery still haunt Edinburgh, whilst in North Wales Wrexham has a number of such relics – notably the Island Green Brewery and the Mount Street premises of Border Breweries, only recently taken over and closed down by Marston's. But perhaps the greatest prize amongst these brewery ghosts, albeit on a much smaller scale, lies behind the Golden Lion in Southwick, Hampshire. Now a museum, with the horizontal steam-engine and mechanical mash tun restored to working order, the Golden Lion has not produced beer commercially since 1956. Yet the brewery and its equipment were left intact, a miraculous survival well worth seeing, not least as a representative of the hundreds of similar brewhouses which have suffered total destruction.

Inside the brewery: the ingredients of beer

The traditional ingredients of beer – and in some places the only legal constituents of a pint – are malted barley and hops, together with water, yeast and possibly brewing-sugars. West Germany's sixteenth-century *Reinheitsgebot* (Purity Law), which is still strictly enforced, restricts the raw materials of German beer to hops, barley and water, while on the Isle of Man the Pure Beer Act of 1874 applies the same restrictions to the island's two surviving brewers, Okell's and Castletown.

On mainland Britain the list of permitted ingredients has steadily grown. There were stiff penalties for those caught adulterating fourteenth-century ale with anything other than water, malt and yeast, and although hops became accepted over the next century there were still pockets of resistance to their use: as late as 1519, as we have seen, Shrewsbury brewers were forbidden to add hops to the brew. Brewing-sugars were first permitted on a regular basis as late as 1847. Present-day brewers have a much freer hand, some of them resorting to cheaper substitutes to make up part of the fermentable material, and the average grist lowered into the mash tun now contains only about seventy-five per cent malted barley; the remaining quarter is made up of such strange additions as flaked maize, wheat flour and even rice or potato flour.

Nevertheless, malted barley universally remains the most important source of fermentable sugars for brewing. Such is its importance that many brewing companies also carried out malting (the conversion of barley grains into fermentable malt) themselves, often at maltings adjacent to the brewery. Brakspear's built an imposing three-storey maltings on the opposite side of New Street to their Henley brewhouse in 1899, whereas Donnington Brewery's malthouse, where barley from the surrounding fields was malted until the 1960s, still forms part of the idyllic group of Cotswold stone brewery buildings. At Belhaven, too, malting was once very important – indeed, until the early 1970s the company's main trade was in malt.

The second of the brewer's raw materials, hops, is now recognized as an essential ingredient of beer despite its relatively late introduction to Britain. After harvesting from the hop yards or gardens of Kent, Hereford and Hampshire, the hops are kilned in oast-houses, where they are spread to a depth of six inches or so and then dried, before being compressed into the hessian pockets in which they may be stored for up to a year.

Third on the list of ingredients is brewing-water, traditionally known in the brewery as 'liquor'. Though it is often taken for granted, it does in fact have a vitally important role to play in forming the character of the beer. The growth of Burton into a major brewing centre was based on the existence of deposits of gypsum under the town: all the Burton brewers (and there were twenty-six of them in 1900) drew their brewing-water from wells giving them supplies of pure water which had percolated through the gypsum and in the process become ideal for brewing pale ales. Marston's, indeed, still draw water from the Field Well at the brewery.

Many of the principal London brewers, unable to brew an acceptable light beer in the capital, where the water was more suited to mild ales and porter, established breweries at Burton in the nineteenth century simply to gain the benefits of Burton water, and one northern brewery, Magee Marshall of Bolton, actually shipped well water from Burton to Bolton in rail tankers. Now, however, the advantage is not so great, for water can be 'Burtonized' by the addition or subtraction of salts.

Yeast, a living single-celled fungus, hardly qualifies as an 'ingredient', yet this more than anything has a crucial influence on the success of the brew. In some cases the same strain of yeast, which multiplies some five or six times with each fermentation, has been employed for decades, as at Samuel Smith's, where the strain dates back to the 1890s. In other cases the yeast is imported on a regular basis, as at Felinfoel, which uses yeast produced by Buckley's, whose yeast strain has survived more than fifty years.

The brewer's day: the process of brewing
In the majority of traditional breweries, which were built on the tower principle, the process of brewing begins at the top of the brewery, and the brew travels downwards by gravity through the successive stages of mashing, boiling and separation from the spent hops, usually being raised back to the top of the brewery at this stage of the process by pumping and then undergoing cooling, fermentation, conditioning and racking. Up to the start of the conditioning process, traditional and keg beers are handled in roughly the same way, but keg beers undergo the extra processes of chilling, filtration and pasteurization, which are designed to kill off the remaining living organisms in the beer – together with most of the flavour. Traditional, living beers undergo a slow secondary fermentation in the cask, imparting subtle flavour characteristics.

The first process at the brewery itself is milling, the preparation of the malted barley for brewing. The malt is hoisted to the top of the brewery and then screened or sieved to grade the individual corns of malt by size; next it is milled to convert it into a coarse powder which is described as 'grist'. After milling, the grist is loaded into the grist case and left to await the start of the mashing operation. The grist is mixed in carefully calculated proportions with the liquor, which has been pre-heated to about 65°C, in a vessel called the mash tun. The hot liquor extracts the fermentable sugars which contribute strength and flavour to the finished product, achieving this in about an hour, and the mash is then allowed to stand for a further period. The sweet malt sugar solution, now known as wort, is then run off from the mash tun through the underback, where the spent grains are retained, to the copper. The spent grains are sparged – sprayed with hot liquor – to remove any remaining fermentable material, and are then removed, to be sold to farmers as cattle feed.

In the copper the hops are added to the sweet wort, and the mixture is boiled, a process which achieves a number of objectives. First and foremost it allows essential oils and flavours to be extracted from the hops, but it also sterilizes the wort and removes undesirable materials left over from the mashing operation. The length of the boil varies from one to two hours, with hops being added gradually and a small quantity being held back until the final stages of the boil, so that their aroma is retained in the product instead of being evaporated in the copper.

The first coppers were open vats heated by coal or wood fires around their base, creating a permanent smokescreen in the brewery. Later, more efficient enclosed coppers were introduced, fitted with a chimney to allow excess steam to escape from the brewery. These coppers were heated by steam passing through coils in the lower part of the vessel; originally they were actually made of copper but latterly stainless steel has become a more popular material. Lorimer & Clark are unique in retaining their open coal-fired coppers, one of them an original dating from the first days of brewing at the Caledonian Brewery, in 1869.

One further ingredient may be added to the mixture which is boiled in the copper, namely brewing-sugar. Although a number of breweries, such as St Austell, still pride themselves on producing beer from malt and hops only, the majority now increase the fermentable material in the brew by introducing sugars into the wort, a practice first permitted on a large scale as recently as 1847. And brewing-sugars are added to the hopped wort in both large and small breweries – from Bass in Burton at one extreme to Wadworth's and Hook Norton at or near the other, for example.

On the completion of the boil two alternative methods are now available to separate the spent hops from the wort. Traditionally the hopped wort is allowed to flow into the hop back, where the spent hops, trapped on the bottom of the vessel, act as a filter through which the wort passes before being pumped up to the cooler. Alternatively the spent hops are separated from the wort by centrifugal force in a whirlpool, which has been installed by many breweries which have updated their facilities recently – notably Gale's in 1983 and both Buckley's and Mansfield Brewery in 1984. The spent hops are removed and then – as with the other raw materials, which all have an after-use – stored and sold, this time for use in the manufacture of fertilizers.

Traditionally the hot hopped wort was pumped into shallow, open troughs located beneath the rafters of the brewery, and allowed to cool down gradually. This was a time-consuming process, however, and it also exposed the wort to bacterial

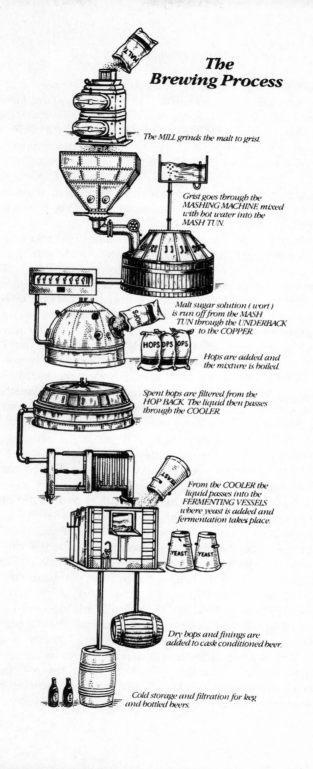

infection, so more modern techniques have been widely introduced. A handful of breweries do, however, still use open coolers in conjunction with the more modern apparatus. Examples are Hook Norton, the Old Swan (Ma Pardoe's) and also Gale's, where the wort flows slowly across a six-inch-deep copper-lined cooler in a room ventilated by wooden-slatted windows. And Elgood's, deep in the Fens, have two shallow cooling dishes, measuring twenty feet by twelve, at the top of their elegant Georgian brewery; the airflow, as at Gale's, is controlled by slatted windows.

Normally the more modern cooling equipment takes the form of one or more heat-exchangers, known as paraflow refrigerators, which cool the beer to about 15°C by pumping it against a counter-current of cold water separated from the wort by the thinnest of metal plates. At the same time the beer may be aerated (that is to say, its oxygen content may be increased) so that a rapid fermentation may begin immediately after the yeast is added.

In one or two breweries, such as Wadworth's and Cameron's, the hopped wort is run from the cooler into collecting vessels, where yeast is added to it and the excise officer is able to measure the quantity and strength of the wort and hence calculate the duty payable on the brew. After between twelve and twenty-four hours the wort is then dropped into the fermenting vessels, a process which stimulates yeast activity by causing aeration. Other brewers, however, run the hopped wort from the cooler straight into fermenting vessels, have the brew 'dipped' for excise purposes and then pitch the yeast.

Now the stage is set for the most critical part of the brewing process, namely fermentation, which consists of the conversion of the wort sugars into alcohol and natural carbon dioxide through the action of the yeast. The yeast is 'pitched' or poured into the wort either in powdered form or as a slurry, and within twenty-four hours it multiplies to form an extraordinary thick, fluffy head or crust on the surface of the wort.

Since yeast, as noted above, typically multiplies five or six times during the early part of fermentation, it is necessary to remove the excess yeast head by skimming, either manually, scooping the yeast off with giant paddles, or by a variety of automatic methods – the James Paine Brewery, for instance, uses a parachute device which draws the excess yeast into an outsize funnel. A proportion of this excess yeast is reserved for pitching future brews, but most is pressed and then sold to yeast-extract manufacturers such as Marmite.

At first the fermentation progresses quickly, with the rapid conversion of a high proportion of the fermentable sugars into alcohol, but as the specific gravity falls, the pace of fermentation becomes more sluggish, and the slowly maturing brew spends some five to seven days in the fermenters. The 'green' beer is then cooled to allow the yeast particles which have become suspended in the liquid to settle out, prior to the commencement of the conditioning process.

Quite a number of different types of fermenting vessels have been used, though originally they were open rectangular or square vessels made of wood, as is still the case in Harvey's fermenting room in Lewes, for example. Later slate was quite often used – Mansfield Brewery brew their solitary traditional beer, 4XXXX bitter, in slate squares, Darley's retain their relatively ancient slate Yorkshire squares because of their effect on the flavour of the beers, which are certainly distinctive, and Samuel Smith's, proud of their status as Yorkshire's oldest brewery, also use slate Yorkshire squares in order to produce a vigorous early fermentation. Boddingtons', however, replaced their slate fermenters with aluminium vessels in the 1920s, mainly because the slate squares suffered from scaling.

Nowadays copper fermenters are commonly in use, and stainless steel, which is easier to clean and maintain, is increasingly popular. Quite often the more modern fermenting vessels are enclosed, as at Hardy's & Hanson's in Kimberley, a brewery which exemplifies the process of gradual expansion and modernization. Until the 1950s slate squares were used, but these were then progressively replaced with stainless steel vessels with twice the capacity, and in 1980 two 350-barrel stainless steel enclosed fermenters were installed to boost capacity further.

One very important variant method of fermenting beer, now unique to Marston's of Burton-on-Trent, is the 'Burton Union' system. Developed in the nineteenth century primarily because it was found particularly effective in conjunction with the yeast normally used in Burton, the system was widely adopted and even spread as far as the Isle of Man, where Clinch's Brewery (now sadly defunct, though the former brewhouse still adorns the North Quay in Douglas) had a set of unions. For many years, however, there were just two breweries using the system, both in

The former Union Room at the Bass Brewery, Burton-on-Trent

Burton – Bass Worthington and Marston's. Bass, however, took the extraordinary decision, primarily at the behest of their accountants, to declare their awe-inspiring union room redundant in the early 1980s, with disastrous consequences for the flavour of Draught Bass.

Marston's, therefore, own the only surviving union room, and despite the higher costs of cleaning and maintenance they have pledged to retain it, because of the superb quality of draught Pedigree, the sole beer produced wholly in this way. Union sets work as follows. The wort is allowed to ferment for a day or so in ordinary open fermenters, before being run into long rows of unions, vast oak casks capable of holding 144 gallons. The action of the yeast causes vigorously fermenting wort to rise with the yeast through swan-necked pipes into the barm trough, a long, open trough which runs above and between parallel rows of unions. The yeast eventually sediments in this trough, while the wort runs back into the unions, only to rise again in due course. The unique flavour of union-fermented beer is said to result from the continual circulation of the yeast through the fermenting wort.

A somewhat similar system was that of carriage casks, which

became obsolete in the 1950s. At Bateman's, where the carriage casks were replaced in 1953, the wort flowed into very large casks suspended above long, open troughs. Yeast and wort which overflowed from the casks during the vigorous early stages of fermentation had to be returned manually to the casks every three hours, day and night, and it was this extra (and expensive) labour, together with much higher losses of wort during fermentation – up to eight per cent – which finally persuaded the Lincolnshire brewers to commission a stainless steel fermenting room.

A much more recent innovation, and one which is still gaining in popularity, is the conical fermenter, a totally enclosed tall cylindrical tank bearing a rather distant resemblance to a space rocket. The advantages are extreme efficiency in yeast activity, with fermentation time reduced to as little as two days, a considerable saving in space, and greater ease of operation, since the yeast sediments in the conical base and can quickly be removed. Brewery modernization now includes the installation of conical fermenters as a matter of course – as at Thwaites in Blackburn, Young's in Wandsworth and Davenport's in Birmingham, for instance. But Guernsey Brewery holds the record for Britain's smallest conical fermenter, a tiny vessel which stands in the brewery yard and is used to brew lager.

There is nothing traditional about the external appearance of conical fermenters, and consequently there are those who would rather praise beers produced in slate squares or Burton Unions, yet there is no doubt that perfectly acceptable traditional draught beer can be produced in these vessels: Thwaites' excellent beers prove the point. The only pre-requisite is that 'batch' fermentation is practised rather than 'continuous' fermentation, where the vessels are not cleaned and emptied between successive brews, and the wort is subjected to high concentrations of yeast.

At the end of the fermentation process the green beer begins a period of conditioning which prepares it for its journey from brewery to pub or club. For traditional beer this conditioning takes place either wholly in the cask or partly in maturation tanks and partly in casks. The second of these options is increasingly popular and has the merit of extracting a higher proportion of the sediment in the beer before it leaves the brewery, hence reducing the time taken in the pub cellar for the beer to drop bright and become ready for serving. At some point, however, the beer will reach the cask – either from the maturation tanks or more directly from the fermentation vessels via a racking back, a vessel designed to control the flow of beer as it is racked into casks. After each cask is filled, it is rolled into the brewery cellars and stored while a slow secondary fermentation, promoted by the remaining yeast still in suspension in the beer, takes place, producing the natural carbon dioxide which will ensure that the beer is in good condition when served.

The casks can be made of wood or metal; opinion varies as to whether the material makes any difference to the beer, though it is probably fair to say that the majority think not. Breweries are still taking the decision to phase out wooden casks – Burtonwood Brewery as recently as 1984, for example – but at the same time others recognize their value, with St Austell using about fifty per cent wood, John Willie Lees racking about nine-tenths of their beer into wooden casks, Wadworth's firmly committed to wood, and Samuel Smith actually delaying the launch of a new beer until sufficient extra wooden casks could be found.

The size of cask will vary according to the level of trade in a particular pub, though those most commonly in use are firkins (holding nine gallons), kilderkins or kils (eighteen gallons) and barrels (often wrongly used as the generic name but actually a cask of a specific size, thirty-six gallons). At either extreme there are additional sizes tailored to particular situations: pins (4½ gallons) are still used in low-barrelage country pubs by some brewers, while hogsheads (fifty-four gallons) are used for fast-selling beers in larger pubs, for example by Holt's in Manchester.

Just before the beer is racked into casks, it may also be dry hopped, that is, a small quantity of hops may be added to impart extra aroma when the beer is served. Usually only the finest hops are used in dry hopping: two ounces of Kentish Goldings per barrel is a standard addition, though Fuller's use three times this hopping rate with the strong bitter ESB. Eldridge Pope, St Austell and Harvey's are amongst other brewers who dry hop their beers, largely to improve aroma but also to convey an impression of extra bitterness.

Another common addition to the beer at this stage is that of priming-sugars, whose function is partly to increase the strength of the beer marginally but more importantly to help to produce the correct level of carbon dioxide in the beer. Some brewers prime all their beers, while others, including King & Barnes, add primings only to their darker, sweeter beers. Finally, either in the maturation tank or in the cask, finings may be added. Manufactured from the swim bladders of certain fish, finings have the effect of clearing residual yeast cells and other solid particles from the beer, depositing them as a thin sediment on the bottom of the cask.

The serried ranks of casks now mature in the brewery cellar until, ten days or more after it began life in the mash tun, the beer is judged ready to be loaded onto the delivery vehicle, known as the dray. Traditionally this was horse-drawn, and indeed quite a few breweries now have teams of shirehorses back in service, some only for ceremonial duties but others pulling drays on a regular basis to pubs very close to the brewery. Vaux in Sunderland, Young's in Wandsworth and Wadworth's in Devizes all use horse-drawn drays on local journeys for economic reasons. However, most beer is now delivered by lorry or even by articulated truck, which may well be temperature controlled.

At the pub the casks are manhandled into the cellar and then have to be stillaged – placed in the position they will occupy while the beer settles (drops bright) and is served: any movement will disturb the sediment and result in cloudy beer and (Britain being populated by people who drink with their eyes) dissatisfied customers. A slow secondary fermentation is achieved by allowing the beer to 'breathe' through a spile peg which is knocked through the shive hole midway along the length of the cask.

Unless the beer is served by gravity, direct from the cask, a method which these days is confined largely to country pubs where the casks are stillaged behind the bar, there are two standard methods of serving traditional beer: handpumps and electric pumps.

The invention of handpumps, beer engines operated by manual pumps, has been credited to Joseph Bramah (perhaps better known as the inventor of an improved water closet), who patented a rather complicated device in 1797. Now beer engines are much simpler, consisting essentially of a suction pump which is designed to allow a single pull on the handle to deliver half a pint from the cask through the pipe into the glass.

Electric pumps and their counter mountings come in many shapes and sizes, and sometimes the latter are visually identical to top-pressure taps, so that care is needed in seeking out traditional beer served in this way. Some breweries, mainly

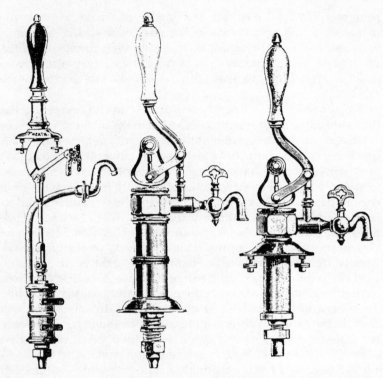

Beer engine fittings from the 1899 catalogue of T. Heath

northern, prefer electric pumps, which use exactly the same basic principle as handpumps but which can operate with smaller-bore pipes (so that less beer stands in the pipe, possibly becoming stale and warm) and can dispense exact, metered quantities of beer. Hence Banks's and Hanson's insist on them in their tied houses, and others such as Hyde's and Ward's use them widely.

In Scotland air pressure or even water pressure is used in the operation of tall fonts which dispense traditional beer; the time-honoured system north of the border, it is gradually being joined by handpumps, which more obviously signify the presence of real ale. Other oddities, in increasing order of harmfulness to the beer, include the cask breather, where a demand valve tops up the level of carbon dioxide in the cask as necessary (possibly justifiable in pubs with very slow sales); blanket pressure, where low-pressure CO_2 prevents contact between air and beer (almost undetectable in a full cask, but the last few pints from a cask can be appallingly gassy), and top pressure, where a much higher level of CO_2 is used to force the beer to the bar (totally unacceptable).

Local brews: who drinks what
Three trends in contemporary beer-drinking are discernible: traditional beers are growing at the expense of their processed counterparts; bitter beers are growing at the expense of mild ales, once the staple drink of the pub-goer but now very much a minority taste except in certain strongholds such as the Black Country; and lager is growing at the expense of everything else.

Mild ales accounted for over forty per cent of draught beer production as recently as 1960 but by 1975 had declined to less than fifteen per cent and now represent less than a tenth of production. Bitter is more difficult to measure (the Brewers' Society figures rather curiously lump together light mild and stout with bitter) but after increasing in popularity in the 1960s and early 1970s, to about seventy per cent of draught production in 1973, it has contracted to less than fifty-five per cent. The reason for this is the growth in lager sales, from three per cent of the market in 1967 to a quarter in 1977, a third in 1984 and projections of forty, fifty and even sixty per cent in the 1990s.

The decline in mild, especially in the south, is a sad story of falling demand and lack of promotion. Fine milds such as Fuller's Hock and Elgood's mild are no longer available in cask form, while brewers such as Ridley's and King & Barnes have amalgamated their mild with ordinary bitter, darkening a small proportion of the brew as necessary, in a last-ditch bid to continue the mild tradition. Yet there are several parts of the country where mild is still the most popular drink: Banks's and Hanson's, the two Wolverhampton and Dudley breweries, produce nearly three times as much mild as bitter (and mild sales are still rising), and in north-west England the practice of brewing two milds has only recently begun to die out (Hyde's and Thwaites are the honourable exceptions, still producing two different milds). In areas such as Wiltshire and Dorset mild has all but disappeared, yet it was once popular here as elsewhere, and widely acknowledged as a cheap and tasy alternative to bitter. Mild enthusiasts would do well to sample some of the excellent beers still on the market – Bateman's DM, Robinson's and Thwaites' best milds and the legendary Brain's Dark, for example – before even more of them disappear.

Bitter, in all its various shades from the yellowish hue of Batham's and Theakston's to the dark reddish-brown of Morland's best bitter and Ruddle's County, is in considerably less danger of extinction. The huge variety still available ranges from the sweet brews of the Midlands, such as Kimberley bitter and Holden's, to more generously hopped (and perhaps dry-hopped) ales, including Holt's remarkable bitter and those of Shepherd Neame and Harvey's. And on another variable, strength, the range extends from the 'boy's bitters' of the south-west, notably St Austell and Devenish, through a plethora of medium-strength brews to the full-flavoured premium bitters, of which Fuller's ESB, Ruddle's County and Greene King Abbot have the greatest reputations and Royal Oak from Eldridge Pope and Arkell's Kingsdown Ale are amongst the most under-rated. Outstanding premium bitters at a slightly lower gravity include Marston's Pedigree and Brain's SA. Strength is by no means everything, however: some of the best beers in the country, beautifully balanced but with a hoppy taste and aroma unsullied by excessive sweetness, are the ordinary bitters from brewers such as Wadworth's, King & Barnes and Bateman's.

In many areas of Britain, especially during the winter months, the beer-drinker's choice is substantially increased by the availability of draught old ales, strong ales and barley wines. Old ale, so called because it is usually matured for longer periods than ordinary beers, is sometimes classed as a strong mild (as in the case of King & Barnes' superb example of the style) and sometimes is not too different except in colour from strong bitters, as with Taylor's rather sweet Ram Tam. Amongst the best are Brakspear's XXXX, Gale's Winter Ale and Young's Winter Warmer. The closely related stock ale, a dark but well-hopped beer unusual in draught form nowadays, is represented by Shepherd Neame's excellent brew. Amongst the draught barley wines, which seem to be slowly increasing in number, the connoisseur would probably gravitate towards Morrell's College Ale, Marston's Owd Rodger and Robinson's Old Tom, whilst the strong ales include the incomparable Wadworth's Old Timer.

Reference to original gravity is commonplace in later chapters, since it is a good guide to strength and, indirectly, value for money. The original gravity of a beer is essentially a measure of the amount of raw materials used in brewing the beer, as a proportion of the amount of water used. So a beer with an original gravity of 1045 would comprise 45 parts of fermentable material to 1000 parts of water. As a rough guide, the minimum gravity of contemporary beers is about 1030, ordinary bitters lie within the range 1035–1040, premium bitters up to about 1050, and barley wines 1070 to 1080. Brewers are now obliged to declare the original gravities of their beers, though the style of their compliance varies from enthusiastic, mainly from brewers of good-value beers of generous gravity, to reluctant, with small notices well hidden from the public gaze.

The bottling department
Bottled beers are in decline: fewer people drink them in the pubs, and even fewer at home, where the can is king. The result is that many local brewers have closed down their bottling lines, often when the equipment needed repair. Some still brew the beers but have arranged for them to be bottled elsewhere under contract. Joseph Holt, for example, trunk beer for bottling down from Manchester to Wolverhampton. In other cases the bottled beers have been lost forever when the bottling hall ceased to function: a particularly sad loss was Brakspear's range of bottled beers, notably the delicious Henley Strong Ale, in 1983.

This reduction of choice in bottled beers will come as a blow to Britain's burgeoning band of collectors of 'breweriana' – advertising-material ranging from old bottles and waiter trays to labels and beermats. Labelled bottles have been in general use since the 1850s, and both the bottles (particularly older screw-stopper types embossed with the brewery's name) and the labels have become eagerly sought after. Full bottles, too, are increasingly regarded as collectable, especially when they contain commemorative brews – Coronation, Jubilee or Royal Wedding ales, for instance, or brews celebrating more local events, such as the Cup Final Ale brewed by Arkell's of Swindon.

But some distinctive bottled beers still exist and, even better, some of them are naturally conditioned in the bottle (that is to say, they contain yeast and therefore undergo a secondary fermentation akin to that of draught beer). All bottled beers were originally produced in this way, but as a result a good many hazy, cloudy or even yeasty glasses of beer must have been served, and the advent of pasteurization was welcomed as offering the possibility of stable, sterile, clear beers with a longer shelf-life. The vast majority of beer bottles today, therefore, are filled with chilled and filtered beers and are then crown corked, pasteurized, labelled and loaded into crates.

Bottling at the Wrexham Brewery in the 1890s

But bottled beers do exist which are not so treated and which therefore have a real, distinctive and mature flavour to them. The best known is bottled Guinness, the staple diet of many drinkers when the keg tide swept the land, and there are also two naturally conditioned beers from the Big Six: Courage Imperial Russian Stout and Worthington White Shield. But the smaller brewers also produce a couple of these extra-special bottled beers, namely Eldridge Pope's Thomas Hardy Ale, the strongest beer brewed in Britain, and Gale's Prize Old Ale, a remarkable barley wine, some of which is filled into old-fashioned corked bottles.

4   South-East England

Astonishingly few breweries survive in the densely populated South-East. Amongst the major Big Six closures since the 1970s are Charrington's Anchor Brewery in Mile End, brewing since 1766 but closed in 1975, and the Courage brewery at Horselydown, established in 1789 and extinct by 1981. Following the closure of Charrington's, Bass now have no brewery (discounting the lager factory at Alton) closer to London than Birmingham – hence, perhaps, the even blander taste of Charrington IPA. And Allied's traditional beer has a good way to travel, from Burton-on-Trent, since Romford has ceased to brew cask-conditioned ale; so Taylor Walker, Benskin's, Friary Meux and the rest of the revived names all get their draught bitter (coincidentally all of the same original gravity, 1037) from the same brewery. The same story is true of Courage since their Reading brewery can produce only processed beer, and hence traditional bitter and Director's Bitter have to be brought all the way from Bristol. The reasons for the ridiculously high prices suffered by London drinkers become somewhat clearer.

Two Big Six brewers do provide traditional beer from local breweries – Watney's from Mortlake and the Truman brewery in Brick Lane (though the best bet in Watney's houses is Ruddle's County), and Whitbread from Fremlin's in Kent and, best of all, Wethered's in Marlow. Although it has not been in control of its own destiny since 1949, when it was taken over by Strong's of Romsey (themselves acquired by Whitbread in 1968), the Marlow brewery continues to produce excellent traditional beer, with the strong and quite dark ale, Winter Royal, an outstanding brew. Other Whitbread breweries in the region have been less fortunate – amongst others, Fremlin's Maidstone brewery has been demolished (together with the breweries of Mason's, Isherwood's and Style & Winch in this once-famous brewing town).

Independent breweries in the South-East are thin on the ground, but the survivors produce some of the best beer in the country. Kent has Shepherd Neame, Sussex has Harvey's (try the outstanding BB) and King & Barnes, Hampshire has Gale's,

London itself the excellent Young's and Fuller's, and in an arc north of the capital there are Brakspear's Henley Brewery, Charles Wells, McMullen's and Ridley's, whose bitter is especially recommended. Many other independent brewers are now represented in the extensive (and lucrative) London free trade, and a significant number, including Vaux, Greene King, Everard's and Thwaites, have bought tied houses in the capital.

The free trade also supports a number of the newer brewers, together with recently established home-brew pubs, especially in London and along the south coast, where names such as Raven and Becket's are prominent. So great is the choice from firmly established brewers, however, that many will be content to limit themselves to the tried and tested brews available in the area. But the choice *is* restricted in one way, for there is virtually no traditional mild available, Fuller's and Young's having been forced by falling demand to stop producing it and others such as Ridley's and Shepherd Neame brewing only tiny quantities of draught mild.

BRAKSPEAR *(Henley Brewery)*

W. H. Brakspear & Sons plc, The Brewery, New Street, Henley-on-Thames, Oxfordshire.
Pubs: 128. Extensive free trade in the Thames Valley and London.
Beers: XXX Mild (1030), PA (1035), Special Bitter or SBA (1043), XXXX Old Ale (1043).

Robert Brakspear, founder of the present company, now very well known for its cask-conditioned Henley Brewery ales, joined one of Henley's firms of common brewers in 1779, but it is worth tracing the history of this firm slightly further back. Until 1768 the Bell Street brewery was in the hands of the Brooks family, but in that year James Brooks and Richard Hayward agreed to become 'co-partners and joint traders in the trade, art and mystery of a brewer'. Hayward was Robert Brakspear's uncle, and in 1779, after Brooks had been bought out, Brakspear joined the brewery, becoming a partner with a quarter share in the business two years later.

Robert Brakspear became the sole owner of the brewery in 1803 and continued to manage it until 1812, the year of his death, when he arranged a merger with the rival New Street Brewery. The combined company had forty-four tied houses but Robert's son,

William, was only ten years old and it was not until 1826 that the Brakspear family reasserted control over the business, William eventually becoming the sole proprietor again in 1848. He remained in control for twenty-one years, then took his sons Archibald and George into partnership, and the business became known as W. H. Brakspear & Sons.

Brakspear's became a public company in 1896, primarily as a means of raising cash in order to buy out the rival Grey's Brewery, and the decline of the family's control started here, accelerating after 1903 when day-to-day management was assigned to two new directors. All but five of the original 8,000 shares were issued to Archibald and George Brakspear, but after the merger 1,500 new shares were issued to Captain Steward of the Grey's Brewery, and he then became the first outside director of the firm. The combined company had about 150 pubs, twenty or so more than today, and two breweries, although Grey's Friday Street premises were quickly shut down and sold.

The Wokingham brewery and its nine pubs were purchased in 1913, and in 1930 eight more pubs were acquired when Wells' Wallingford brewery merged with Ushers. Two years earlier it had, ironically, been the opposition of the 'outside' rather than the family directors which had defeated a proposal to merge with Wethered's of Marlow, a move which would certainly have seen the end of the Henley brewery. The final acquisition, in 1941, was that of Gundry's Goring Brewery, with eighteen tied houses. These extra pubs would have brought the total to about 185 had it not been for the selective closure of uneconomic houses, a process which continues today and which has so far accounted for about sixty pubs.

In 1915 the Brakspear family still owned sixty per cent of the 9,900 shares, with other directors holding twenty-five per cent and outsiders fifteen per cent. The family continued to hold a controlling interest, refusing tempting offers at a time when local competitors Wethered's, Nicholson's of Maidenhead and Simond's of Reading were taken over, until the early 1960s, when a speculator threatened to bid for the company, close the brewery and sell off the pubs. In self-defence Brakspear's approached Whitbread, and it was arranged that the latter would buy a protective stake – currently around twenty-seven per cent. Nowadays the family own a similar share of the company, other directors and staff rather less, and outsiders (of which the largest is the Sun Alliance insurance company, with seven per cent) the remainder.

The beers which have survived because of Whitbread's 'umbrella' are sold in 126 tied pubs (the number is slowly decreasing as small rural houses without modern facilities are sold without licence, amid complaints that Brakspear's are oblivious to the resulting damage to village communities) and a substantial free trade, accounting for about one-third of output, within about fifty miles of the brewery. There are two magnificent draught bitters, PA (1035) and Special Bitter (1043), which despite their different strengths share the same dry, hoppy, distinctive character. The Old Ale (1043), available in the winter only, is another fine drink, mellow and warming, while the mild (1030) is thin and rather disappointing compared with its stablemates.

Three-quarters of Brakspear's production is real draught beer, and there is a heavy emphasis on tradition, with superb beers in unspoilt country pubs, many of them hidden in the Chiltern woodlands. But the company recognizes the need to keep up-to-date, too, running its own computer company and prepared to take harsh decisions – closing the bottling line in 1983, for example, because of reduced demand, despite redundancies and the loss of respected beers such as Henley Strong Ale. And there has been recognition that the layout of the brewery site is hopelessly inadequate for modern conditions, having been designed in the horse-and-cart era; hence Brakspear's have entered the planning minefield, pointing out that, 'To remain in business the company must either update its facilities or cease to exist.' With luck it should never come to that.

FULLER'S
Fuller, Smith & Turner plc, Griffin Brewery, Chiswick, London W4.
Pubs: 135. Widespread free trade. Beers: Chiswick Bitter (1035), London Pride (1041), ESB (1056).

Fuller's beers are certainly very good at gaining awards, ESB and London Pride between them having picked up the 'Beer of the Year' award at four of the first eight CAMRA Great British Beer Festivals, and at venues as far apart as London, Leeds and Brighton. Yet it has taken an extraordinarily long time for the brewery to install traditional beer in the majority of its own pubs, and even now about a quarter of Fuller's pubs have bright beer only. Real mild, too, has been withdrawn in recent years because of falling sales.

The three beers which do survive in traditional form include one which spent the 1970s as a chilled, filtered and carbonated beer but was reintroduced as a true draught beer in 1980. This is Chiswick Bitter (1035), a pleasant if unexceptional ordinary bitter. Much better known is London Pride (1041), an excellent, fruity, individual premium bitter which was originally known more parochially as Chiswick Pride. A stronger cousin of this fine ale is Extra Special Bitter (1056), the strongest bitter regularly produced by an established brewery in Britain. It is based on the same recipe as 'Pride' but is quite heavily hopped (and dry hopped, with six ounces of Goldings per barrel), giving what the head brewer describes as a 'heavy, aromatic, meaty flavour'.

The origins of ESB, however, are of special interest, for this is a successor to a strong, dark mild, Old Burton Extra, which declined in popularity in the 1960s and was relaunched (and strengthened) as Winter Bitter in 1969. Two years later Winter Bitter, which had had disappointing sales, became ESB, and in 1974 it suddenly caught the drinking public's imagination – sales increased sixfold, and it now accounts for a respectable minority of production. Sadly, Old Burton Extra's historic lower-gravity counterpart, Fuller's Hock, is now a thin bright-only beer, its 300-year reign as one of Britain's best dark mild ales having come to an end in 1979, by which time it accounted for less than two per cent of output despite special attempts to popularize it.

Traditional Hock was a link, albeit tenuous, with the brewery's earliest days, around 1660; a generation later the Mawson family acquired the brewhouse on Chiswick Mall, and it was only in 1782 that their grip was broken, though the consequences of this change in ownership were almost catastrophic. The Thompsons were the next to exercise some degree of control, though in the 1820s 'financial difficulties caused by deceit and mutual distrust' between two Thompson brothers indirectly led to the first Fuller involvement, when John Fuller of Neston Park, Wiltshire, became a partner in 1829. Worse was to follow, for the Fuller/Thompson partnership had to be dissolved in 1841 when Douglas Thompson fled to France to escape his creditors. Four years later John Bird Fuller poached Henry Smith of Ind & Smith, the Romford brewers (later Ind Coope), together with his brewer John Turner, and the present firm was born.

Today the much enlarged company, which has rarely grown by acquisition (Brentford's Beehive Brewery in 1910 is an exception) but more commonly has picked up individual pubs to extend its estate, is still largely run by Fullers and Turners. The great-great-great-grandson of John Fuller is the chairman, and a new generation of Turners became directors in 1985. Expansion is very much in the air: a fifty per cent increase in capacity achieved in 1981 has quickly proved insufficient, and in 1986 the company raised more than £1 million by the issue of new debenture stock to finance further growth. Fuller's look an unstoppable force in independent brewing – even the thirteen per cent Whitbread stake, less publicized than many of the 'umbrella' shareholdings, is discounted as a threat – and even those who are surprised by the accolades their beers have received in recent years will hope that the Griffin Brewery, which has been producing good beer since Cromwellian times, continues to prosper.

GALE'S
George Gale & Co Ltd, The Brewery, Horndean, Portsmouth, Hampshire.
Pubs: 96. Beers: XXXL Light Mild (1030), XXXD Dark Mild (1032), BBB (1037), Winter Brew (1044), HSB (1051).

Gale's, with its south-coast location and wide range of cask beers, was well placed to take advantage of the boom in real ale in London and the South-East in the 1970s, yet failed to gain more than a toehold in the free trade. However, this was due largely to lack of production capacity, and a major brewery expansion completed in 1984 has seen the company pursuing a more aggressive policy, looking for new outlets and gaining a very useful contract to supply its premium bitter, HSB, to many Phoenix Brewery (Watney's) houses in Hampshire and Sussex.

The Gale family is still very much involved, even though the majority shareholding passed to the Bowyers, millers and maltsters from Guildford, as long ago as 1896. Now the majority of the shares are held in family trusts, and both Bowyers and Gales are represented on the board of directors. In 1847 Richard Gale acquired a former home-brew pub, the Ship and Bell in Horndean, together with the brewery on the opposite side of the road, and his son George rapidly developed the business despite a fire which destroyed the original brewery. The fine Victorian brewhouse which stands next to the Ship and Bell was built in 1869, and by 1884 trade had expanded to include Portsmouth and Southsea. The present company was registered four years later.

In the early twentieth century Gale's embarked on a quite vigorous strategy of expansion by take-over; had it not petered out in the 1920s, the company might by now have been a leading regional brewer. The first victims were Clarke's Homewell Brewery in Havant in 1903, adding nine pubs to Gale's estate; in 1907 the Square Brewery in Petersfield was acquired, and in 1912 at least one pub was bought when the Wickham Brewery ceased brewing (some of the brewery buildings survive behind the King's Head, now a Gale's pub). Finally the Angel Steam Brewery in Midhurst, with five pubs, was taken over in 1923 and closed down after four years.

Since then progress has been steady but unspectacular, with the occasional purchase of single pubs but no major expenditure. Now there are ninety-six pubs, all with traditional beer, in an area stretching from Romsey to Brighton, and from south Oxfordshire to the Isle of Wight. Until recently the brewhouse was largely unchanged from 1869, with an original hot liquor tank, a nineteenth-century malt screen, and wooden fermenting vessels more than half a century old. The copper, however, dates only

Prize Old Ale – one of the few naturally conditioned bottled beers

from the 1940s, having been installed as a matter of necessity – its predecessor exploded! The shallow cooling tray in the eaves is also an unusual survival. Since 1984, however, a brand-new range of equipment has increased capacity by forty per cent, at a cost of more than £600,000.

Two of Gale's beers in particular have gained wide reputations in recent years. The most unusual is Prize Old Ale (1095), a rich and sweet naturally conditioned barley wine brewed at the same gravity since it was introduced in the 1920s, and matured in hogsheads for several months before being hand-filled into half-pint bottles which are in many cases corked rather than crown-corked and still have raised glass lettering. It would be a tragedy if declining sales of bottled beers, or the need for expenditure on the bottling plant, hastened the end of this unique ale. The second of these brews of renown is HSB – Horndean Special Bitter (1051), first brewed in October 1959 at a slightly higher gravity and modelled on Bass Pale Ale. Its original selling price was 1s.6d. a pint. Now it is widely respected as a potent, warming and full-bodied best bitter, much stronger and tastier than present-day Bass; indeed a number of breweries have in turn tried to reproduce its bitter-sweet flavour.

The staple brew for many Gale's drinkers is the lighter, hoppier and no less distinctive BBB (1037), a slightly variable brew but nevertheless very popular. It also acts as the base for Winter Brew (1044), which is a mix of BBB and Prize Old Ale; a dark and sweet winter warmer which has an unexpectedly strong hop flavour, it is brewed from October to March. Finally, there are still two milds, though the future of the dark mild (1032), a thin but pleasant brew which is in only a few pubs, is uncertain; XXXL light mild (1030), despite its low gravity, is a well-balanced beer with a rather wider sale.

HARVEY'S
Harvey & Son (Lewes) Ltd, Bridge Wharf Brewery, Lewes, East Sussex.
Pubs: 30. Expanding free trade, now including London. Beers: XX Mild (1030), PA (1033), Sussex Best Bitter (1040), Old Ale (1043).

More than almost any other brewery company in the country, Harvey's signify the attraction and also the vitality of traditional values in the brewing industry. They are small but ambitious, with recent additions to the tied estate and a major expansion of

Harvey's Brewery in Sussex, from across the River Ouse

Ruddle's Brewery in Rutland in the early twentieth century – then owned by H. H. Parry

Theakston's Brewery in Caldewgate, Carlisle – **former** home of the State Management Scheme

Jennings Brewery and Cockermouth Castle in Cumbria, from across the River Derwent

The cask yard and malthouse at Morland's in Oxfordshire in the 1920s

King & Barnes Brewery, Sussex, in 1900

The bottling hall at Davenport's, Birmingham, in the 1930s

The coppers at Davenport's Brewery, Birmingham

The mash tuns at the United Breweries, Abingdon, Oxfordshire, in the 1920s

The top of Davenport's conical fermenters

Burtonwood's cooper at work before his retirement in 1984

the brewery; they brew excellent draught beers, increasingly widely available in the free trade; and, perhaps most important of all, they seem reasonably well protected against an unwelcome take-over bid, with the shares still tightly held by descendants of the founder (though with no one controlling interest) and a very clear determination to retain their independence.

The seventh generation of John Harvey's descendants is now actively involved with the company, which has seen a number of major changes since he established the Bridge Wharf Brewery by the River Ouse in Cliffe High Street around 1790. Early growth was slow, but by 1881 the volume of trade justified a complete rebuilding (sparing only a small fragment of the original Georgian structure), which was carried out to the design of William Bradford, a noted architect who also designed the Hook Norton Brewery in Oxfordshire. The similarities between the two breweries, exuberant Victorian Gothic in style, are obvious.

The second major change took place in 1938, when Tony Jenner was recruited as head brewer. Under his guidance a rather moribund country brewery was transformed into the present company. He remains today as chairman, and his son has followed him into brewing at Harvey's. The third change, twenty years after the arrival of the Jenners, saw the taking of a bold decision. By this time Lewes, home to six or seven breweries for much of the nineteenth century, was now host to only two: Harvey's and Beard's. The latter, whose Star Lane brewery operated from cramped premises and had suffered from yeast infection, decided to cease brewing and have since obtained their beer from Harvey's, a vitally important boost to trade which still survives, although Beard's – a completely separate company – also buy in certain beers from other brewers.

The final significant development is the doubling of mashing capacity, from 25,000 to 50,000 barrels a year, with the addition of a second mash tun, in a slightly smaller tower in front of Bradford's original, and in an almost identical design. New fermenting vessels may well be required as demand continues to grow, but the new mash tun means that peak demand, which has previously outstripped supply in August, can now be met.

The demand is very largely for BB, now better known as Sussex Best Bitter (1040), which accounts for more than three-quarters of draught beer sales. This is a magnificent beer, one of the truly great and distinctive bitters which are still available; quite sharp to the palate but nevertheless essentially malty in character, it is regarded as well suited to local tastes and so, very reassuringly, there are no plans to follow other, more short-sighted breweries by reducing its distinctive nature. The beer is, surprisingly, also available in keg form, although only a handful of kegs are filled each week. The other bitter, PA (1033), suffers by comparison with its illustrious partner but is nevertheless a finely hopped light bitter first brewed during the Second World War.

Harvey's also offer two darker beers in traditional draught form. XX Mild (1030) is an unremarkable dark mild for which demand has steadily declined, so that only eight barrels a week are now brewed; its future must be perilously uncertain. In total contrast, sales of the winter-only Old Ale (1043) are buoyant, and its reappearance is eagerly awaited each year. Brewed to a recipe for a pre-First World War mild ale, with high proportions of crystal malt and black sugars, Harvey's Old is a marvellously tasty, nutty beer which is well worth searching out.

One final draught beer is Elizabethan (1090), a very strong pale barley wine originally brewed for the Coronation in 1953. It is very rarely available on draught but is also worth trying in bottle, together with the equally strong Christmas Ale, which is much darker and has a very high hopping rate. Harvey's bottled beers, which together account for only a tenth of total output, also include Sweet Sussex, a very sweet stout, and Blue Label, a dry hopped pale ale.

KING & BARNES

King & Barnes Ltd, 18 Bishopric, Horsham, West Sussex.
Pubs: 56. Significant free trade. Beers: Sussex Mild (1034), Sussex Bitter (1034), Old Ale (1046), Draught Festive (1050).

While King & Barnes attribute their recent growth in large measure to the revival in demand for cask-conditioned beers, they have not been slow to seize upon other opportunities in the 1980s. In particular, they launched their own JK lager in 1985, replacing the Bass and Watney brands which had grown to encompass forty per cent of their tied house sales. The brewery, too, has seen major development, with an entirely new brewhouse erected, a quality-control laboratory established and computerized systems installed in the offices.

Both the King and the Barnes branches of the present family-controlled company entered the brewing business in Horsham,

A beer mat from King & Barnes of Sussex

buying separate breweries which had existed since about 1800. James King began as a maltster in the Bishopric, eventually amalgamating with Satchell's North Parade brewery in 1870. Within a few years brewing had been transferred to the Bishopric and James King had bought out the Satchells, the business now being styled King & Sons. Meanwhile the Barnes family had purchased Usher, Robins & Company's East Street Brewery in 1878. Their independence was short-lived, however, for 1906 saw the creation of King & Barnes Limited and the closure of the East Street brewery; all operations were centred on the Bishopric, on the site still referred to as the North Brewery (the title of 'Horsham Brewery' could not be adopted until the sole remaining local rivals, Michell's West Street Brewery, closed in 1912).

The new company consolidated rather than expanded and by the 1960s were short of cash and apparently ripe for take-over. As

John King, a former chairman, candidly said in 1978, 'Twenty years ago we just couldn't afford to go over to keg. Now we just smile and say we would never have dreamed of doing it.' Take-over was resisted and the 1970s saw the real-ale revival and a vast increase in production (which more than doubled between 1975 and 1985), with free trade growing to form two-fifths of output. A completely new brewhouse, with three times the capacity of the original Victorian version, was completed in 1980, allowing the premium Festive bitter to be produced in draught form for the first time. The brewery buildings are now modern, functional and rather unlovely but the process inside the new brewhouse is just the same, as the beers testify.

The new brewhouse is used for brewing Sussex Bitter and Festive, the adjacent old brewhouse for the smaller brews of mild, old ale and bottled stout. Sussex Bitter (1034) is the staple diet of many local drinkers, rightly so since it is quite possibly the best low-gravity bitter in the country. Beautifully balanced but with a very distinct smack of hops (the beer is dry hopped, and the hops are from choicest Kent and Worcester sources), it is a delightful, aromatic beer. Draught Festive (1050) has not achieved quite the same reputation since its launch in 1980, but it is a full-flavoured premium bitter which is popular in the free trade. Sussex Mild (1034) is not a separate brew these days, being simply the bitter darkened with caramel, but despite its four-degree rise in gravity in 1980, when the previous recipe was abandoned, it has failed to generate volume sales. Old Ale (1046) is an exceptionally tasty winter brew, which the brewery regard as a strong dark mild.

There are some especially fine country pubs in the fifty-six-strong tied estate (the Blue Ship near Billingshurst and the George & Dragon at Dragon's Green are two that spring to mind), and in 1981 King & Barnes introduced their 'Ale Trail' as a means of publicizing them. This passport scheme offers prizes of sweatshirts or engraved glasses to drinkers who have a pint in all the King & Barnes pubs, and several hundred enthusiasts have already completed the trail at least once. Many of them will have seen another of the brewery's recent introductions, a horse-drawn dray which since 1984 has been making local deliveries; this is, of course, a re-introduction, since the original drays and horses were sold as long ago as 1919.

McMULLEN'S

McMullen & Sons Ltd, The Hertford Brewery, Hartham Lane, Hertford.

Pubs: 160. Free trade in Essex, Herts and London. Beers: AK Light Mild (1033), Country Bitter (1041), Christmas Ale (1070).

McMullen's, the sole survivors of the thirty-five Hertfordshire brewers in business in 1904, have in their midst another remarkable survivor – AK, a best mild first brewed late in the nineteenth century, and one which (remarkably, given its trading area in south-east England, where many draught milds have recently bitten the dust) still sells more than the bitter.

McMullen's Brewery in Hartham Lane, completed in 1893

The precise date of origin of AK is as uncertain as the meaning of its initials. Some of the brewery's literature gives the date as 1829, only two years after the brewery was founded, but the first documentary evidence (a price list in which it is offered at a shilling per gallon) dates from only 1888, and the truth probably lies somewhere between the two, with some time in the third quarter of the nineteenth century most likely. As for the intitials, the K may be an old mark denoting a light beer (Greene King and Ind Coope both used KK for their light milds), but the A has never satisfactorily been explained. They combine as the name of one of the finest light milds in the country, a quite well-hopped 1033 brew which is as full of flavour as many ordinary bitters. Indeed, the brewery themselves have sometimes been unsure what to call the beer in the past: a pump clip designed in 1965 described it as 'AK mild bitter'!

AK has two traditional draught colleagues, Country Bitter (1041), a malty and distinctive best bitter, and Christmas Ale (1070), which is a strong seasonal ale with a good reputation. Together with McMullen's lagers, a premium beer called Steingold and the low-gravity Hartsman, introduced in 1984, they are produced in two adjacent brewhouses. One of these is a well-known Hertford landmark, a classic Victorian tower brewery erected in 1891, whilst the other was completed in 1984. Despite its relative youth, this is a brewhouse largely of traditional design, allowing the existing brewing process (which gives short shrift to unmalted barley or pelleted hops, for example) to continue unaltered; there is a good deal of lager-processing plant, however, for it was the decision to brew an ordinary lager instead of relying on Harp which precipitated the expansion.

The 1891 brewery was the culmination of the firm's early growth, from its origins in 1827 when Peter McMullen brewed the first pint in Railway Street, Hertford. The founder bought his first pub, at Bengeo, in 1836, and as business grew he transferred to a brewery at Mill Bridge. His sons continued the expansion, acquiring the Cannon and Star breweries in Ware in 1864 and 1874 respectively, and building the new brewery in Hartham Lane. By 1898, when the Waltham Abbey Brewery was bought, McMullen's owned more than ninety pubs, and this number grew further with the acquisition of the Epping Brewery (1907) and the Hope Brewery in Hertford in 1922. The company is still family-controlled, with the great-great-grandsons of the founder in

SOUTH-EAST ENGLAND

A McMullen's poster of 1931

charge today, and the shares tightly held and cleverly organized so as to minimize the risk of take-over.

McMullen's are one of the few breweries to have invested heavily in bottled beers recently, opening a brand-new bottling hall in 1979 and installing new plant in 1980 to centralize beer and mineral-water production and allow the introduction of a range of 'own-label' fruit juices. Amongst the bottled beers are Mac's No. 1, a pale ale which is also sold as a keg bitter, and Castle Special Pale Ale. But well over half of production is accounted for by AK and Country Bitter, beers which have in the past too often been spoiled by McMullen's insistence on blanket or top pressure to prevent air coming into contact with the beer but which are now sold by handpump in three-quarters of the 160 houses, many of which provide notably comfortable surroundings.

RIDLEY'S
T. D. Ridley & Sons Ltd, Hartford End, Chelmsford, Essex.
Pubs: 65. Some local free trade. Beers: XXX Mild (1034), PA (1034), HE (1045), Bishop's Ale (1080).

It is the opinion of many connoisseurs that Ridley's brew some of the finest cask-conditioned beers available in Britain. It is also very likely that the average drinker has never heard of them, for despite their proximity to London Ridley's never seem to have been unduly tempted by the free trade, with the result that the beers have to be sought mainly in the company's pubs, a good proportion of which are unspoilt rural locals, often serving the beer direct from the cask.

The Ridley family became involved in brewing in about 1840, having previously concentrated on milling and malting; and today the family still have control of the brewery's destiny. Thomas Dixon Ridley was the instigator, shortly after 1800, and by 1840 the company owned two cornmills on the River Chelmer, at Hartford End and Felsted. Shortly thereafter the Ridleys diversified into malting, acquiring several maltings in northern Essex, and brewing. The brewing operations were concentrated in the tiny village of Hartford End, a few miles north of Chelmsford, and amongst Ridley's advantages were the facility of drawing excellent well water from beneath the brewery, and the opportunity of using their own locally grown barley, which they themselves had malted.

Despite these advantages, growth has been gradual and unspectacular, although by 1906, when the present company was incorporated, there were forty-seven Ridley's pubs, together with the Hartford End brewery and four maltings, at Writtle, Felsted, Chelmsford and Braintree. Subsequently the number of public houses has grown slowly to sixty-five, many of them acquired individually, although one or two tiny breweries have been taken over, notably Charles Brown & Sons of Hatfield Peverel in 1939. Conversely, the flourmilling and malting operations have been discontinued, the former in 1958 and the latter (which supplied not only Ridley's malt but also some for the major brewers) in the early 1970s.

The beers for which Ridley's are renowned, perhaps partly because they are stored in wooden casks, are PA, a fine sharp bitter and (a relatively recent introduction) HE, a deceptively powerful premium bitter. Sadly, Ridley's draught mild is to all intents and purposes no longer with us, though a beer does exist which is described as XXX mild. Until the late 1970s the mild was a delicious, creamy and only slightly sweet ale with an original gravity of 1030; now it is identical to the PA except for the addition of caramel to make it acceptably dark. There can, however, be no argument with the brewery's decision to effect this change, for the mild accounts for a mere two per cent of production; nevertheless, a distinctive traditional beer has been lost.

The ordinary bitter, PA (1034), is in no such danger. This is an excellent, delicate brew, well hopped but with a malty aftertaste, and light enough to be entertained as a 'session' beer. The same cannot be said of HE (1045), a sweeter and much stronger bitter which was introduced into eight pubs in 1983 and has quickly become much more widely available. The initials are those of the brewery village, Hartford End. The final traditional ale is Bishop's Ale (1080), a really rich, fruity and powerful barley wine which is not often found on draught.

About a third of Ridley's production is destined for bottling, a quite astonishingly high proportion in view of the dramatic national decline in the popularity of bottled beer. One reason for this is the quality of the beers, which in addition to Bishop's Ale include stout, unremarkable light and brown ales, Stock Ale (an unusual brew which is well worth trying) and Old Bob, a strong pale ale. The second reason is that Ridley's have brewed for Cook & Sons of Halstead, an unusual firm with no pubs but a number of off-licences, since the latter closed their Tidings Hill Brewery in 1974.

Ridley's also supply beer to pubs owned by, and formerly supplied by, Gray's of Chelmsford, another firm who ceased brewing in 1974 and now buy in their beers, notably from Greene King. And a third venture which Ridley's have found profitable is a beer swap with Truman, which has seen their draught beer in a number of Truman pubs in return for keg Ben Truman in Ridley's pubs. Given these trading agreements, it is perhaps not surprising that Ridley's have chosen to remain within a relatively restricted area, supplying excellent beer at extremely low prices (by south-eastern standards) and yet making profits which mark them out as one of the most efficient brewers in the country.

SHEPHERD NEAME
Shepherd Neame Ltd, Faversham Brewery, 17 Court Street, Faversham, Kent.
Pubs: 250. Free trade accounts for a high proportion of production. Beers: Master Brew Mild (1031), Master Brew Bitter (1036), Stock Ale (1037), Invicta Best Bitter (1044).

For some time Shepherd Neame have been the sole remaining

independent brewers in Kent, a county which once boasted brewing towns such as Maidstone and firms such as Fremlin's – arguably Kent's premier brewers before they succumbed to the advances of Whitbread. Now 'Sheps' is the local brew, despite the continuing presence of Whitbread Fremlins at the adjacent ex-Beer & Rigden's brewery in Faversham, and it is a brew which has been noticeably improved in consistency and palate over the last twenty years without the loss of its highly individual hoppy character.

Faversham, ideally situated to use Kentish hops and barley, and with a copious supply of fine brewing-water, has been producing beer since the Cluniac abbey was founded by King Stephen in 1147. All trace of the abbey has been lost, although Shepherd Neame have used its existence in marketing their keg bitter and other products. Brewing on the present site is of fairly ancient origin, too, Richard Marsh having established a brewhouse here in 1698. The first Shepherd, Samuel Shepherd of Deal, bought the brewery from Marsh's widow in 1741, although early expansion was modest and when Julius Shepherd inherited the brewery from Samuel late in the eighteenth century there were only three tied houses – the Castle, Three Tuns and Red Lion, all in Faversham. Henry Shepherd, Samuel's great-grandson, took Percy Beale Neame into partnership in 1869, and on Henry's death eight years later the Neames became the sole proprietors. Since then the Neames (and the Johnstons, descendants of a brother-in-law of Harry Neame) have been firmly in control.

Shepherd Neame has always relied heavily on the free trade. In the early twentieth century club trade in the Kentish coalfield accounted for up to half of production. Gradually a more stable tied-house base has been built up, however, in parallel with an ambitious thirteen-year brewery modernization and expansion programme. Previously the only large-scale injection of tied pubs followed the acquisition of Mason's Waterside Brewery in Maidstone, with about eighty houses. Recent expansion has been on a piecemeal basis, with sixty-six pubs acquired in the 1970s from Big Six brewers who, with their different accounting perspectives (and less distinctive products), found them uneconomic. The tied estate in London has risen from two to more than twenty by these means.

All but a handful of the pubs serve traditional draught beer, with ordinary, or Master Brew, bitter (1036) accounting for two-thirds of demand. This is an extremely well-hopped, dry and clean-tasting bitter with a very distinctive flavour – so much so that there are many Kentish drinkers who will not drink it, presumably because its magnificently bitter, hoppy tang contrasts with the blander bitters in other local pubs. The stronger bitter, Invicta (1044), is of recent origin, having replaced a 1039 brew, Best Bitter, in 1984. Ostensibly this was because 'best' was too similar to the ordinary bitter and was in less than a quarter of the tied houses, but ironically Invicta is scarcely more widely available. The darker beers are mild (1031), thin, dark and sweet, and in some danger since it accounts for less than ten per cent of draught beer production, and Stock Ale (1037), a primed and darkened version of the bitter but nevertheless an unusual and, like the rest of its stablemates, very distinctive ale.

WELLS
Charles Wells Ltd, The Brewery, Havelock Street, Bedford.
Pubs: 278. Beers: Eagle Bitter (1035), Bombadier (1042).

The original Charles Wells bought the long-established Horne Lane Brewery in Bedford in 1876. To celebrate the centenary his descendants closed the cramped town-centre brewery – and moved to a brand-new brewery, very modern and functional in appearance, on a much larger site across town. The brewing equipment, much of it from Germany, includes enclosed conical fermenters and a computerized push-button control panel, and the process of brewing takes place not in mash tuns and coppers but in a mash-mix vessel, lauter tun and whirlpool. Furthermore, because of the long distance between conditioning tanks and racking equipment, Wells remove the yeast by centrifuge in the conditioning area and then replace it when the beer is racked. Doubts have been raised about this process by real-ale diehards, but since the beer does undergo a secondary fermentation in the cask it surely qualifies as 'traditional'.

For Wells do have a traditional image, despite the high-technology aspects of their brewing operation. The brewery is still family-run, with five Wells on the Board, and there are two widely available traditional draught bitters which arouse fierce debate between protagonists and detractors. This is particularly so with Eagle Bitter (1035), formerly known as IPA. Admirers point to its crisp, sharp flavour, hoppy with a malty aftertaste;

less charitable souls consider it thin and unexceptional, and out of its depth in the London free trade. Only a few, perhaps, would regard it as an exceptional ordinary bitter. Bombardier (1042) dates from 1980, when it replaced the ill-fated 1050 brew Wells Fargo, which had irritated the American banking group of the same name but had failed to capture a sufficient slice of the premium bitter market. Bombardier, though, is eminently drinkable, a fine, full-bodied malty bitter.

Much of Wells' production is for the own-label supermarket trade, with Sainsbury and Waitrose amongst their customers, so there is quite an emphasis on processed bitter; there are also three keg bitters, the strongest of them the quaintly named Noggin. A great deal of lager is also brewed for the take-home trade, including the bottom-fermented Kellerbrau and the stronger Red Stripe lager, brewed under licence from Desnoes and Geddes of Jamaica. Originally this was aimed at the West Indian community in Britain, but it is now distributed by wholesalers in Birmingham, Leicester, Nottingham and Cardiff. One other brewing venture should be mentioned, namely the home-brew at the Ancient Druids in Cambridge, for in December 1984 Wells became the first regional brewery to open a home-brew pub. Two beers are brewed, using malt extract rather than a proper full mash: Kite Bitter (1035) and Druids Special (1044).

All this is a far cry from 1876, when the Horne Lane brewery and its thirty-five public and beer houses were offered for auction at the Swan Hotel in Bedford. In the normal course of events Charles Wells would not have been interested, since he was a steamship captain, but his marriage plans were endangered when his prospective father-in-law refused to let his daughter marry a sailor. So, aided by his father, Wells bought the brewery, already established for half a century or more, and began to consolidate, acquiring the nearby Cardington Brewery in 1897 and tapping a supply of good brewing-water by sinking a well in 1902 (the well is still in use, but the water now has to be pumped some two miles to the new brewery). Charles Wells was succeeded as chairman by three of his sons in turn, and during this time two more Bedford breweries were taken over – the Phoenix Brewery (and its subsidiary, the St Paul's Square Brewery) in 1923 and Fuller's in 1935. The final acquisition was the Abington Brewery Company of Northampton and its twenty-three pubs in 1963. More recently individual pubs have been bought, notably in

London, where there are now five Wells houses. There are currently 278 pubs, two-thirds of them with traditional beer.

YOUNG'S
Young & Co's Brewery plc, The Ram Brewery, Wandsworth High Street, London SW18.
Pubs: 145. Beers: Bitter (1036), Special (1046), Winter Warmer (1055).

Young's, who could still be relied upon for a pint of traditional draught beer when the rest of London was awash with Red Barrel, Double Diamond and the other keg concoctions of the 1960s and early 1970s, have had to increase their brewhouse capacity dramatically, at a cost of over £5 million, to keep pace with demand, which is still mainly for their traditional ales (although two 'draught' lagers are also produced these days). As John Young remarked in his chairman's statement in 1982, 'The big brewers are trying to put the clock back and real ale, an unmentionable name in the keg beer age, and for which we were much teased, is now actively promoted. . . . How marvellous to see respect for the return of individual taste and character.'

The brewery contains a fascinating combination of ancient and brand-new vessels, from the steam-driven beam engines (in regular use until the early 1980s and still in working order) installed by Wentworth & Son of Wandsworth in 1835 and 1867, and a copper which has been boiling wort since 1869, to conical fermenters which are used mainly for brewing the two lagers and other processed beers. The new equipment is not just for draught beers, either: in a move which bucked the trend for the industry, Young's ordered a new £1 million bottling line in 1985.

Beer has been brewed on the site of the present brewery, close to the River Wandle and once connected by both river and canal to the Thames, since at least 1675, when the Draper family had control. They were succeeded by the Trittons in 1763, and in 1831 the executors of George Tritton sold the brewery for £140,000 to a partnership of Charles Allen Young and Anthony Fothergill Bainbridge. The brewery took its name, the Ram Brewery, from the Ram Field opposite, possibly once the home of the Wandsworth villagers' communal ram, and today a golden ram tops the brewery's weathervane and a live ram is the brewery mascot. It is by no means the only livestock in Young's farmyard, however: ducks, geese and a goat also have their homes here, and the

A Young's advertisement

famous shirehorses, which have been shown regularly since 1924 but are kept primarily for delivering beer to the forty-seven pubs within three miles of the brewery.

The partnership between the Youngs and the Bainbridges was finally dissolved in 1884, and Charles Young's son now traded under the title 'Young & Co'. The last century has seen the creation of a public company and substantial growth, but the enterprise remains firmly under the control of the Young family and the Ram Brewery Trust, which operates the profit-sharing scheme and other employees' benefits and now owns twenty per cent of the company. Growth has been generally organic – and it still is, with about one pub a year acquired over the past decade, though more pub-buying is planned now that the brewery development is completed. Some acquisitions have, however, taken place, with William Wells' Britannia Brewery in Allen Street, Kensington, taken over in 1924 (the Britannia, still a Young's house, was once the brewery tap) and Foster Probyn, a major beer and mineral-water bottler, acquired in 1962.

Young's traditional beers, justifiably renowned and often rewarded with prizes at Brewex (the international brewers' exhibition), are now three in number, Best Malt Ale having been laid to rest in 1983. This disappointingly thin and bland dark mild accounted for two-fifths of sales just after the Second World War but by 1980 constituted only five per cent of production. A paler replacement, Young John's Ale, lasted only a few months before it, too, was withdrawn. But the three remaining ales are delicious: a noticeably sharp, very hoppy and full-flavoured ordinary bitter (1036); Special Bitter (1046), strong and distinctive, still with a rather hoppy nose; and Winter Warmer (1055), which until the late 1960s was, in common with Fuller's strong ale, known as Burton, and which is justifiably one of the best-known winter ales, a powerful dark old ale which has exactly the effect its name suggests.

5 Southern and South-West England

Breweries in the main production centres of the South-West, such as Bristol, Plymouth and Exeter, proved irresistible to the Big Six, and the major companies there were generally bought up by the 1960s, largely by Courage (who acquired the main Bristol concern, George's, in 1961 and also snapped up the biggest Devon brewers, Plymouth Breweries, for £6.5 million in 1970) and Whitbread (Norman & Pring, Exeter in 1962; Starkey, Knight & Ford of Tiverton in the same year; and, biggest of all, West Country Breweries of Cheltenham, with no fewer than 1,300 pubs, in 1963). Bass owe their considerable presence in the region largely to Charrington's purchase of Brutton Mitchell Toms of Yeovil in 1961, and Watney's still run Wiltshire's largest brewery, Usher's of Trowbridge. Whitbread's Cheltenham plant and Courage's in Bristol are the only other remaining Big Six breweries, apart from small-scale revivals such as Hall's (Allied Breweries) plant in Plympton.

In the remoter and more rural areas there has been a surprisingly high survival rate. Cornwall, remotest of all, has the Blue Anchor home-brew pub in Helston, together with St Austell and Devenish (now brewing only at Redruth, though the main centre of operations until 1985 was Weymouth). Dorset, where six independent breweries existed in 1958, still has three – Eldridge Pope, Hall & Woodhouse and Palmer's; of the others Devenish has migrated rather than been taken over, and local rivals rather than the Big Six have accounted for Matthews & Co of the Wyke Brewery, Gillingham (annexed by Hall & Woodhouse in 1963) and John Groves (bought by their next-door neighbours in Hope Square, Weymouth – Devenish – in 1960). Wiltshire, too, has plenty of local ale, from Arkell's, Gibbs Mew and Wadworth's. The other southern survivors are Morland and Morrell's, in Oxfordshire, and the exquisite Donnington Brewery in Gloucestershire.

There is almost a surfeit of competition from newer independent brewers, especially in the more popular tourist areas such as south Devon. Probably the best known is the Golden Hill Brewery in Wiveliscombe, Somerset, whose Exmoor Ale achieved the rare distinction of collecting a Beer of the Year Award at the Great British Beer Festival in 1980. Amongst the best are Archer's of Swindon, offering a wide range of really excellent ales, and Blackawton Brewery in south Devon, one of the first of the new wave of breweries in Britain, having been established in 1977. Significantly, none of these breweries offers a mild, for again this is bitter-drinking territory, to such an extent that only St Austell of the south-coast brewers, and Donnington, Morland and Morrell's further north, even brew a cask mild. But many brewers offer two or more bitters of varying strength, and there are some excellent strong bitter beers such as Hall & Woodhouse's Tanglefoot, Eldridge Pope's Royal Oak and Arkell's Kingsdown Ale – to say nothing of the extraordinary range of strong brews which can be sampled at the Blue Anchor, the most historic of Britain's home-brew pubs.

ARKELL'S

J. Arkell & Sons Ltd, Kingsdown Brewery, Swindon, Wiltshire.
Pubs: 64. Considerable free trade in Wiltshire, the Thames Valley and London. Beers: John Arkell Bitter (1033), BBB (1038), Kingsdown Ale (1050).

In 1843 John Arkell, a farmer newly returned from Ontario, built a brewery at Lower Stratton, then in the countryside near Swindon. It was a classic case of a farmer progressing from growing barley to malting it and then to brewing. It was a shrewdly timed move, too, for the Great Western Railway was just completing its workshops in Swindon, and demand for Arkell's beer was such that a much larger brewery was built on the family's farmland at Kingsdown in 1861.

After John Arkell's death his sons continued the expansion of the company; the third generation saw the firm into the 1950s; and Peter Arkell of the fourth generation is the present chairman (his cousin Claude is the proprietor of Donnington Brewery in the Cotswolds). The fifth generation of Arkells is already represented on the board, the next is growing up, and since the whole of the shares are family-owned, the brewery's destiny is still entirely in the hands of the founder's descendants – unlike some 'family' companies which have allowed control to spread more widely and have eventually found themselves at the mercy of the financial institutions or, worse still, their competitors.

The draught beers, brewed from a grist containing eighty-five per cent malted barley, are all dry-hopped and as a result have a truly bitter, country flavour to them. Perhaps the best is BBB (1038), widely known in the free trade as a refreshing alternative to some of the heavier, sweeter beers available. John Arkell Bitter (1033) rivals it for popularity in terms of barrelage but is rather less distinctive.

A relatively recent addition to the range of traditional draught beers is Kingsdown Ale (1050), an excellent strong bitter with an interesting history. Introduced in 1978 as a draught strong ale with an OG of 1060, Kingsdown was well received but because of its strength (and, consequently, high price) failed to sell well. In 1981 the gravity was dropped ten degrees, sales doubled and the beer became recognized as a fine, malty, powerful bitter. Its market is still comparatively small, but it appeals particularly to the London free trade.

The fate of darker beers from Arkell's is in direct contrast. Draught mild was long since abandoned and more recently even the bottled brown ale has been lost. Emphasizing the swing to lighter beers, Arkell's also brew North Star keg, an entirely separate brew whose sales are restricted largely to the local club trade, and two lagers. As recently as 1979 Arkell's were quoted as having no intention of brewing their own lager, and their conversion is a telling illustration of the recent growth of the lager trade. The brews are a standard, low-gravity lager and a premium brew called 1843. Arkell's have also lent their name to a home-brew kit, but this is actually made up by Edme in Essex, from a recipe supplied by Arkell's.

The keg and lager beers are all fairly bland, but the two traditional bitters can be outstanding when at their best, and so the gradual conversion of Arkell's pubs from pressure dispense to real ale has been most welcome. Even as recently as 1975 only three houses sold traditional beer, but now the number *without* it is in single figures. Or so it seems: the pubs certainly have cask-conditioned beer but the brewery recommend the use of a light blanket pressure (half a pound of carbon dioxide) to prevent air coming into contact with the beer. Fortunately many landlords recognize that this practice is unnecessary – unless sales are very slow, the beer will have been sold before it can possibly have been contaminated – and happily they neglect to connect the carbon-dioxide cylinder.

The Arkell's tied estate now numbers sixty-four, with a heavy concentration in Swindon and the remaining pubs, with one notable exception, within thirty miles of the brewery. Improvement of the houses has a high priority – in fact, the company has long been known for the quality of its pubs – and a six-figure sum for refurbishment is not unknown. New pubs have been built, too, including the Liden Arms, a fine timbered pub on a Swindon housing estate, and the Woodshaw Inn at Wootton Bassett, opened in 1985.

The exception to the rule regarding pub location is the Duke of Edinburgh at Winkfield, near Ascot, acquired because much of Arkell's trade is now in this area. About forty per cent of production now goes to the free trade, much of it in London and the Thames Valley – including the British Rail bars at Reading and London Paddington.

BLUE ANCHOR
The Blue Anchor Inn, 50 Coinagehall Street, Helston, Cornwall.
Home-brew house. Beers: Mild (1040), Medium Bitter (1050), Best Bitter (1053), Special (1066), Extra Special (1070).

Every one of Britain's four long-established home-brew pubs is, by definition, pretty unusual, but the Blue Anchor is not only the oldest and most difficult to get to, it also boasts the most extensive beer list amongst them – one so extensive, in fact, that it puts to shame many of its commercial cousins. The beers are devastating: heavy, rich brews which can carry a disconcerting yeast haze, they demand respect and have caught out many visiting drinkers in their time.

Initially the visitors were monastic, for in 1400 the Blue Anchor was a monk's rest, an idyllic-sounding place where they could escape to find peace and contemplation, not least through brewing their own ale. By the eighteenth century it had become a fully fledged inn, with an enlarged brewery and a brand-new skittle alley, which became a popular venue for locals attempting to score 'floorers', by knocking all nine pins down with a single ball. The custom was to take beer into the alley by the gallon – at a cost of 1s.4d. for the eight pints. A century later the Cornish tin-miners were paid in the bar of the Blue Anchor, clearly still one of the focal points of the small market town of Helston.

By that time the Blue Anchor had also achieved a degree of notoriety, with a landlord murdered – James Jones, who ill-advisedly intervened in a bar-room brawl in 1790 – and two suicides, one of whom contrived to hang himself in the skittle alley and the other, an even greater eccentric, throwing himself down the twenty-five-foot-deep well at the rear of the inn (the water from the well, incidentally, is still used for brewing).

Perhaps as a result of these events the inn fell into decline, for when it was auctioned in the 1860s there would have been no bids had it not been for Thomas Richards, who called in for a drink and came out having bought the inn and brewery; he had to borrow the deposit and then walk five miles home to raise the purchase price. His son – born in the room above the bar – introduced the present Spingo ale just after the First World War, and his grandson Geoffrey Richards was the brewer and licensee for forty-two years until 1975, since when it has been let on tenancy.

The mash at the Blue Anchor comprises 110 gallons (only about three barrels) but there are four or five brews a week of Spingo in summer, and two new fermenting vessels have had to be added to keep up with demand. Until the late 1970s Spingo was available in three strengths: ordinary (1033), medium (1050) and best (1053). Medium and best survive and are powerful malty brews, pale-coloured and with a yeasty haze at times. Recent changes have seen the demise of ordinary bitter, which was a pretty unusual brew, and the introduction of three new beers, including a very welcome, full-flavoured mild (1040). The others simply extend what was already an adequate range of strong beers, with Special Bitter (1066) and Extraspecial (1070) rich, heavy brews which approach the character of barley wine in some respects. The Blue Anchor itself is comparatively unchanged, however, a superb drinker's pub most easily recognized in Coinagehall Street by its thatched roof. Inside, the skittle alley has been rebuilt, and live jazz has been introduced, but the main attraction without doubt continues to be the home-brewed ales.

DEVENISH
J. A. Devenish plc, Redruth Brewery, Redruth, Cornall.
Pubs: 330. Extensive free trade. Beers: John Devenish Bitter (1032), Cornish Best Bitter (1042), Wessex Best Bitter (1042).

Major reorganization has occurred at Devenish in the 1980s, with rationalization of production and distribution facilities at Redruth and Honiton respectively, and the closure of the Devenish Weymouth Brewery in 1985 – ending a tradition of more than

200 years brewing at Hope Square and shifting the focus of the company's interests dramatically westward, towards Devon and Cornwall. The result of this flurry of activity was a certain amount of take-over speculation, culminating in 1986 with news of an agreed merger between Devenish and Inn Leisure, a relatively young pub-owning company.

The end of the Weymouth brewery was particularly sad, and it was unexpected too, for an earlier reorganization, in autumn 1984, had seen it re-designated as the company's specialist cask-beer brewery, producing ordinary bitter and Wessex Best Bitter only. Within a year, however, the decision had been taken to close Weymouth because of rising costs, the need for investment and the growth in demand for lager, and to transfer all production to Redruth. This spelled the end of one of the oldest breweries in operation in Britain (not continuous operation, though, for the brewery was badly damaged in 1940 during a German attack on the Weymouth naval base, and was out of action for two years, during which time beer was supplied by Eldridge Pope and Groves).

The brewery was being run by the Fowler family in 1742 (brewing having reputedly been carried out on the site since the thirteenth century). In 1824 William Devenish bought the brewery, one of three then operating on Hope Square, and began to expand its area of operations in Dorset and Devon. The twentieth century brought massive expansion as a series of take-overs brought other breweries into the group: these included Carne's Falmouth Brewery in 1921, bringing the first substantial Cornish connection, the Well Park Brewery in Exeter four years later, the Redruth Brewery Company in 1934 and Vallance's of Sidmouth in 1957.

Vallance's Brewery had been established in 1832 by Richard Searle, but the Vallances had gained control within twenty years and held on until Woodhead's South London Brewery bought them out at the end of the Second World War. The London firm clearly found it impossible to manage at such a distance and hence they sold it to Devenish, with thirty-two pubs. Three years later Devenish were buying again, though this time the target was a more local rival, John Groves & Son, whose Hope Brewery adjoined the Devenish Weymouth brewery. Groves ran 120 pubs and also had a depot in Swindon, so that this was by far the biggest fish landed by Devenish. There were now more than 400

Devenish pubs, but the figure has been reduced substantially in recent years – the policy, much criticized though understandable, is one of 'recycling capital from unprofitable properties to ones providing better opportunities'. Low-volume pubs are thus still being sold – seven in 1984 alone, with another fifteen up for sale during the next twelve months.

Devenish no longer brew cask-conditioned mild, having withdrawn the fruity, sweetish XXX Mild, brewed at Redruth, in 1983. The ordinary bitter, John Devenish Bitter (1032), brewed from malt, hops and sugar only, is similar in palate to a light mild, however, and sales of this are buoyant. Wessex Best Bitter and Cornish Best Bitter (1042) are more robust and are dry-hopped; 'Wessex', though, now has a 180-mile journey back to the Weymouth pubs. Traditional beers are now normally served direct from the cask or by handpump, a great improvement on the situation in 1970, when top-pressure dispense was the rule, mainly because of the seasonal variations in trade.

DONNINGTON BREWERY
L. C. Arkell, Donnington Brewery, Stow-on-the-Wold, Gloucestershire.
Pubs: 17. Some free trade. Beers: XXX Mild (1034), BB (1034), SBA (1041).

Donnington Brewery is a gloriously unexpected survival in this age of take-overs and brewery closures. Not a limited company, nor even a partnership, its fate rests in the hands of one man, Claude Arkell, cousin of the chairman of Arkell's of Swindon. It has further remarkable features, too, including a magnificent lakeside setting in the Cotswold countryside and outstanding traditional beers served in all seventeen pubs.

The history of the brewery dates back to 1865, but the brewery buildings pre-date their present use by several centuries. They began life as Donnington Mill, in the manor of Broadwell, in the thirteenth century, and three centuries later were in use as a clothmill. In 1580 this was rebuilt and converted into two cornmills, which were bought by Thomas Arkell in 1827. Finally, in 1865, Richard Arkell, the grandfather of the present proprietor, began brewing here.

Disaster nearly struck when Richard Arkell's three sons quarrelled for years over their inheritance; the dispute was finally settled when John Arkell bought out his brothers. His contribution towards the business was to build up the tied estate of idyllic, stone-built Cotswold country pubs, which numbered sixteen by the time his son Claude took over in 1952. By 1955 one has been sold and two bought, and those seventeen have formed the estate since then. The only cloud on the horizon is the succession, for Claude Arkell has no children, and although he takes a philosophical attitude ('I have made no provision for the

Donnington Brewery beer and stopper labels from the 1940s

future . . . the future is in the hands of the Almighty'), there must be a real doubt over the survival of the Donnington Brewery at the end of the Arkell reign.

The situation of the brewery is stupendous: the yellow-grey Cotswold stone of the brewery buildings and the fine brewery house complement the millpond, with its wildfowl and fish farm. The pond is the source of the River Dikler, a tributary of the Windrush, and even that noble Cotswold river has nothing to match the comprehensive beauty of the Donnington scene. The brewery itself has seen little change to its outward appearance over the years, though much of the equipment inside has been updated and the buildings now house an odd mixture of old and new. The older elements include the millwheel, providing waterpower for a variety of pumps and engines, an open copper still in regular use, and an ancient chiller, bought from the ill-fated Blatch's Theale brewery in Berkshire. Newer elements include the stainless steel fermenting vessels and a pasteurizer, recently added to give the bottled beers extra shelf life.

The draught beers, which are all dry-hopped, are XXX Mild (1034), a bland dark ale which is brewed once a month and supplied to the few pubs which number mild-drinking Midlanders amongst their customers (it is a tribute to Claude Arkell's consideration for his customers that he has kept faith with the mild this long), BB (1034), a delightfully fresh and hoppy bitter which accounts for two-thirds of Donnington's output, and SBA (1041), a much heavier-drinking brew which is a really outstanding sweet, dark and rich bitter. The bottled equivalents are brown ale, light ale and Double Donn, but demand for them is now so low that bottling takes place only twice a month.

The ingredients are spring water, drawn from a strong natural spring next to the millpond, malt (which until about 1960 was milled from corn grown on the surrounding farmland, much of it in the hands of the brewery), local Worcestershire hops and a small amount of brewing-sugar. Adjuncts such as flaked maize and wheat flour have been tried but rejected as inferior.

All the seventeen pubs offer the draught beers in their traditional form, although this has not always been the case, and it was only in 1977 that representations from drinkers were heeded and carbon-dioxide pressure was removed from the beer in ten of the pubs. Until then the brewery had argued that it was needed to keep the beers in condition, but now demand is such that this is unnecessary. Outstanding amongst the pubs are the Fox, idyllically situated at Broadwell and renowned for its food; the Mount at Stanton, right at the top of the village and with a selection of gardens; and the Halfway House at Kineton, successfully catering for a wide range of customers in a single bar.

ELDRIDGE POPE

Eldridge Pope & Co, plc, Dorchester Brewery, Dorchester, Dorset. Pubs: 180. Beers: Dorchester Bitter (1033), Dorset Original IPA (1041), Royal Oak (1048), together with Thomas Hardy Ale (1125), a naturally conditioned bottled beer.

One of the most forward-looking of the local and regional brewers, Eldridge Pope brew excellent traditional beer but sadly make it available in traditional form in only a minority of their houses. Indeed, they go so far as to make a virtue of this lack of reliance on traditional beer, Christopher Pope remarking in 1984 that, 'Several years ago when our friends in the wooden cask brigade of country brewers were smiling from the headlines, we were perfecting our production facilities and our lager, adventuring in and improving our pubs, and consolidating our wine; now we find events moving towards us.' Certainly their Faust lager, brewed under licence from the small Bavarian brewery of that name, is regarded as noticeably superior to most British lagers.

The company is dominated by Popes, with five of them on the seven-strong board (and the other two directors, one of whom is also chairman of Palmer's, are cousins of the Popes!). Yet it was Charles Eldridge who began the business, taking over the Antelope Hotel in Dorchester with his wife Sarah in 1833 and establishing the Green Dragon Brewery four years later. After Charles' death Sarah Eldridge carried on in partnership with Alfred Mason, until Edwin Pope acquired Mason's share in 1870. Within four years Edwin and his brother Alfred had acquired complete control of the business; though this control has lessened, the board, including Alfred Pope's grandson and three great-grandsons, still speaks for more than a quarter of the voting shares.

By 1880 the firm had outgrown the Green Dragon Brewery, and a four-acre site on the other side of Dorchester, close to the main railway line, was purchased from the Duchy of Cornwall. The splendid Victorian brewhouse was badly damaged by fire in 1922 but restored within two years; more recently plant for lager brewing and conditioning has been added alongside it. In the early years Eldridge Pope concentrated their hunt for tied houses along the line of the London to Bournemouth railway, and in the process they took over a number of smaller firms in Dorset, including Styring's Poole Brewery in 1900, the licensed houses of Frampton's Steam Brewery in Christchurch in 1920 (the brewery itself soldiered on, supplying private customers, for another fourteen years), Woolmington's of Sherborne in 1922 and, in an excursion into Hampshire, Young's Brewery in Twyford and its seven pubs, bought at auction in 1911.

The 180 or so pubs are now dispersed over a very wide area, with a handful in London and others in Bristol, Bath and Exeter, but the main concentration is still in rural Dorset and Hampshire. A good deal of time and money has been lavished on the pubs, with many of them offering high-class food and others specially designed to attract whole families. It is a pity that so few of them serve the beers free of top pressure, blanket pressure or the cask-breather device, all of which can affect the taste – and that they do so despite Christopher Pope's stated preference for traditional beer without carbon dioxide.

Of the three traditional draught beers, Dorchester Bitter (1033) is the largest seller, perhaps surprisingly since, although it is a refreshingly bitter brew, it is relatively weak and of no great character. Dorset Original IPA (1041) is an excellent and very well-hopped bitter, but the outstanding Eldridge Pope draught beer is Royal Oak (1048), dry-hopped and hence surprisingly bitter to the taste but at the same time full-bodied and malty. The beer was introduced in 1974 and is based on the recipe for a draught pale ale brewed in 1896. Even this is eclipsed, however, by Thomas Hardy Ale (1125), first brewed in 1968 for a Thomas Hardy festival and still brewed as near as possible to his description of Dorchester strong ale in *The Trumpet-Major*: 'the most beautiful colour that the eye of an artist in beer could desire; full in body, yet brisk as a volcano; piquant, yet without twang; luminous as an autumn sunset; free from streakiness of taste, but finally rather heady'. Hardy Ale is naturally conditioned and will improve in the bottle for at least ten years, and it will keep for twenty-five, as befits the strongest beer brewed in Britain.

GIBBS MEW

Gibbs Mew plc, Anchor Brewery, Salisbury, Wiltshire.
Pubs: 70. Beers: Wiltshire Traditional Bitter (1036), Premium Bitter (1042), Salisbury Best Bitter (1042), Bishop's Tipple (1066).

Gibbs Mew is a small company with a fairly unusual recent history. It was one of the first local brewers to abandon traditional draught beer in the 1960s (even today there are still five keg beers), and hence it was left without decent beer when the real-ale movement gathered momentum; a draught barley wine plugged that gap in 1976. Their acquisitions, too, have perhaps been more trouble than they were worth: a venture with a former clubs' brewery in Lancashire lasted just two years, and the London wholesaling business of Robert Porter, acquired in 1979, sustained heavy losses before it was sold in 1985.

The company dates back to 1898, when the assets of two Salisbury brewers, Bridger Gibbs & Sons of the Anchor Brewery and Herbert Mew & Company of Castle Street, were merged. The Gibbs family have been brewing for over 200 years, and their interests were first consolidated when George Bridger Gibbs set up in partnership with Sydney Fawcett in Endless Street, Salisbury, in 1838. George's nephew Bridger Gibbs was employed here before moving first to the Bell & Crown, Catherine Street (still a Gibbs' house), and then, in 1858, to the present site, taking over the long-established Anchor Brewery.

Three acquisitions stand out in the history of Gibbs Mew. The first, that of soft-drinks manufacturers William Seymour & Co (Sherborne) Limited, was uncomplicated and the subsidiary company still makes a useful contribution to profits. But in 1962 Gibbs Mew, by now committed to the keg revolution, bought the Clough Springs Brewery in Barrowford, near Burnley, formerly the home of the Lancashire Clubs Federation Brewery, which had ceased trading in 1960. The idea was to produce keg beers for the club trade in northern England, but within two years the venture had been abandoned because competition from other brewers allied to high transport costs from Salisbury had rendered it uneconomic.

The third acquisition was that of Robert Porter, the East London wholesaling and former beer-bottling business best known for its bottled Guinness and Bulldog Brand pale ale. Bought in 1979 for £900,000, the loss-making firm was brought into profit within twelve months, but then disaster struck. According to the chairman's report in 1982 'ineffective control organization together with some deliberate deceit as massive pilfering and mistakes were allowed to go ahead' cost at least £100,000 in 1982 and the Porter subsidiary made a £470,000 loss. In 1984 the loss was a 'quite unacceptable' £615,000. No wonder the business was sold – to the James Paine brewery, for an estimated £½ million, in March 1985.

Despite these potentially disastrous diversifications, Gibbs Mew continue to brew at Salisbury and, since the Gibbs family still owns comfortably more than half the share capital, there is no reason why this state of affairs should not continue.

Real ales have flowed out of the Anchor Brewery in increasing numbers since 1976, when Bishop's Tipple (1066) was first introduced as a strong, sweet, peculiarly distinctive strong ale. Second in line was Premium Bitter (originally 1039, now 1042), which in its early days was essentially Anchor keg bitter without the processing. In 1977 Gibbs even produced a draught Jubilee Ale to celebrate the Queen's Silver Jubilee. More recently 3X mild has been tried unsuccessfully, but Wiltshire Traditional Bitter (1036), originally brewed as a 'foreign' ale for Ansell's (with Ansell's bitter in Gibbs houses in return), and Salisbury Best Bitter (1042) have lasted longer. Premium Bitter is now a brew similar to the last-named; unfortunately none of them is especially noteworthy.

Despite the proliferation of traditional beers over the last few years, Gibbs Mew still relies on keg beers for a good deal of its business, inevitably so since it is heavily committed to free trade in pubs, hotel and clubs throughout southern England. Many of these outlets are unable to look after traditional beer – hence Anchor, Wiltshire Special, SPA, Great Bustard and Super Mild keg beers, the last three with very low original gravities of 1031 or 1032. Drinkers in Gibbs Mew pubs can expect better, however, with traditional beer in most of them and a fair number operating almost as free houses, with a variety of other brewers' ales also on the bar.

HALL AND WOODHOUSE (Badger Beers)

Hall & Woodhouse Ltd, The Brewery, Blandford Forum, Dorset.
Pubs: 160. Very extensive free trade from Devon to London.
Beers: Hector's Bitter (1034), Badger Best Bitter (1041), Tanglefoot (1048).

A celebratory bottled beer from Hall & Woodhouse

A beer mat advertising the strong and hoppy Tanglefoot

Badger Beers, as the Hall & Woodhouse ales are popularly known, have undergone something of a transformation in recent times, with the 'boy's bitter' replaced by the slightly stronger Hector's Bitter, Best Bitter gaining a steadily increasing reputation in the south of England free trade, and a completely new strong bitter called Tanglefoot introduced. A brewery which was little known outside its native county only ten years ago has thereby achieved an unlikely prominence – yet one which has been richly deserved, given the fine and improving quality of its traditional beers.

The founding of the brewery was the work of Charles Hall, who was born into farming stock but chose to open a brewery in the village of Ansty in 1777. The Woodhouse family became involved in 1847, when the founder's son Robert took George Edward Woodhouse (who had married his niece) into partnership, and they soon became the driving force, for Robert, who died in 1858, was the last Hall to play a part in running the business. Today there are three Woodhouse directors, and the shareholdings are still tightly controlled. The company's Badger trademark was first used in 1875 – no records survive to explain the choice – and in 1882 Hector's Brewery in Blandford St Mary was taken over from John Hector's successors, Horace Neame (of the Kentish brewing family) and Thomas Cock, although for many years Ansty remained the headquarters and two breweries were operated.

In the late nineteenth century seven draught beers were on offer – X Ale, XX Mild, XXX, XXXX Old Ale, Invalid Stout, AKA ('made from choicest malt and hops, strongly recommended' – and only 1s. per gallon) and India Pale Ale. In 1898, the year in which the twenty-six pubs of Godwin's Durweston brewery were bought, a limited company was formed. In 1900 it faced a serious challenge when most of the Blandford brewery was destroyed by fire, but fortunately, a new and larger brewery was under construction nearby, at a cost of £28,000. The Fontmell Magna brewery was acquired in 1904, and the Marnhull Brewery eight years later – after all, a larger brewery needed more outlets, as became apparent in the 1960s.

The Badger trademark was not registered after its introduction, and this omission led to a bizarre series of court proceedings instigated by a mineral-water manufacturer who also used a badger trademark, one Joseph Verity of Pateley Bridge in York-

ESTABLISHED — TRADE MARK. — 1777.

HALL & WOODHOUSE,

BREWERS,
WINE & SPIRIT MERCHANTS,
AERATED WATER MANUFACTURERS,

AND AGENTS FOR THE

SCOTTISH AMICABLE LIFE,
Imperial Fire and Accidental Insurance Companies.

ANSTY AND BLANDFORD BREWERIES.

ALES.

		Per Gall.
X Ale...	... ...	-/10
XX Mild	... ...	1/-
XXX	... ...	1/4
XXXX Old October	... ...	1/6
Invalid Stout	... ...	1/6
A.K.A. made from choicest Malt and Hops, strongly recommended		1/-
India Pale Ale		1/4

Malt and Hops at current prices.

ALL KINDS OF WINES AND SPIRITS IN STOCK.

DELIVERED FREE IN CASKS OF 5 GALLONS AND UPWARDS.

Telegrams: "Woodhouse, Ansty, Melcombe Bingham."

A Badger ales price list from the late nineteenth century

shire, at the turn of the century. Quite why he was concerned by Hall & Woodhouse's use of the animal, given the wide separation of the two firms' trading areas and their different products, is unclear, but the courts found in his favour, and it was only in 1929 that Verity agreed to relinquish his rights to the badger for the princely sum of £52.10s.

An unusually protracted take-over involved Matthews' Wyke Brewery in Gillingham, which Hall & Woodhouse valued at £100,000, including its sixty-eight tied houses, in 1930. Negotiations broke down because the proprietor was unwilling to agree that part of the purchase price should be left on mortgage; it was eventually acquired in 1963. By this time Ellis's brewery in Wimborne had been bought (1935) and the original Ansty brewery had been finally closed down (1937), whilst Hall & Woodhouse had failed in their bid to take over Blandford's other brewery, Marsh & Sons, which fell to Simond's of Reading in 1939.

Since the Wyke brewery take-over there have been more than 150 Badger pubs (now including four in London), almost all of them now selling truly distinctive draught beers which are notably well hopped and, incidentally, fermented with a yeast strain which dates back to 1934. Even Brock lager has a considerable pedigree, having first been brewed in 1959. The draught beers are Hector's Bitter (1034), a pleasant hoppy bitter introduced in 1983 to replace the 1031 ordinary bitter, which was an exceptionally well-hopped and tasty low-gravity beer whose passing was mourned by many; Badger Best Bitter (1041), a really outstanding dry, full-flavoured brew; and Tanglefoot (1048), a fairly dark strong bitter which still retains a noticeably hoppy palate.

MORLAND
Morland & Co plc, 40 Ock Street, Abingdon, Oxfordshire.
Pubs: 220. Widespread free trade in the Thames valley. Beers: Mild (1032), Bitter (1035), Best Bitter (1042).

The foundation of Morland, one of the oldest of Britain's independent brewers, can be traced back to 1711, though the present site in Abingdon was acquired somewhat later. The business was originally established in West Ilsley, now an attractive farming and horseracing community high on the Berkshire Downs. John Morland's purchase in 1711 was of a maltings, though his son Benjamin quickly diversified into brewing (the site of the brewery can still be identified). The ambitions of the family finally outgrew the capacity of the country brewery on 11 February 1861, when Edward Henry Morland bought the Eagle Brewery, Abingdon, at a public auction in the Cock and Tree public house in Ock Street, opposite the brewery. To this day the Eagle brewery site remains the centre of the company's operations.

Expansion was rapid in succeeding years, with the Abbey Brewery in Abingdon quickly acquired – brewing in the town had originally been in the hands of the monks, and for some time Morland's stored ale in the Abbey Brewery's historic cellars – and two acquisitions in 1889, the Saxby Brewery in Abingdon and Field & Sons' Shillingford Brewery near Wallingford. Two years earlier the present public company had been established. Until 1913 all deliveries were made by horse-drawn drays (forty-six horses and four mules were needed) but the purchase of two steam wagons in that year began the decline of the traditional dray, except during the Second World War, when the delivery fleet was commandeered by the Army.

A new and much larger brewery was built on the site of the Eagle Brewery in 1912, and the extra capacity was soon utilized to supply the pubs of the Wantage Brewery Company, the Royal Albert Brewery in Reading, Hewitt's Waltham Brewery, the Angel Brewery at Reading and Belcher & Habgood's Tower Brewery in Abingdon – five companies which had been taken over and seen their breweries closed down by 1939. Since then expansion has been limited to the occasional building of new pubs, and the upgrading of the existing tied estate.

Since the 1950s Morland's have sought the protection of the Whitbread 'umbrella', and with some forty per cent of the shares in the hands of the big brewer (and large holdings listed under investment companies) Morland must be considered vulnerable to take-over, or to the kind of deal which saw Border Breweries, another company with a substantial Whitbread interest, taken over and quickly closed down by Marston's.

Morland's beers – which gained early recognition with the award of a silver medal at a national brewer's exhibition as long ago as 1888 – are dry, well-hopped and clean-tasting, held in high regard by many. Their characteristic flavour derives in part from the use of brewing-liquor from wells on the Eagle Brewery site, a yeast strain which has been retained since the turn of the century, barley from the chalk soils of the Berkshire Downs, and hops which contain a proportion from the local hop farm at Kingston Bagpuize.

The draught beers, cask-conditioned but served under top pressure in some outlets and (though handpumped) stored using a cask breather in many others, are XX Mild (1032), a dark and nutty beer which is unusually bitter tasting; PA Bitter (1035), a delightfully light low-gravity bitter with a very distinctive taste; and BB (1042), stronger and slightly sweeter but still with a dry, hoppy flavour. The bottled beers are of unusual interest, too. Old Speckled Hen (1050), strong and malty, commemorates the fiftieth anniversary of MG's presence in Abingdon; the name is derived from the colours of the Featherweight Fabric Saloon, a demonstration model from 1929. Monarch Strong Ale (1065) is an unusually bitter-tasting strong ale.

There are some 220 pubs, many of them close to the brewery in Abingdon, but covering an area from agricultural Faringdon to cosmopolitan Windsor. The pubs' signboards depict George Morland, a distant relation of the early brewers who was a painter of rustic scenes (examples hang in the National Gallery) and who was imprisoned for debt in 1799 and again in 1804, the year in which he died.

The company has followed a policy of investing heavily in the tied estate, a policy which has produced some excellent traditional pubs but which has also, as the chairman has said, led to a situation where, 'Sometimes there have to be casualties, when a house has to be closed because its commercial prospects do not justify the high cost of bringing it up to the standard now demanded.' Morland therefore sides with the big brewers in sometimes closing small country pubs – something which some of its independent competitors still try to avoid at all costs.

MORRELL'S

Morrell's Brewery Ltd, The Lion Brewery, St Thomas's Street, Oxford.
Pubs: 137. Beers: Light Ale (1032), Dark Mild (1033), Bitter (1036), Varsity (1041), Celebration (1066), College Ale (1073).

Mark and James Morrell entered the brewing business in Oxford in November 1782, when they took over the lease of the Lion Brewery from the Christchurch Estate. Brewing had, however, been taking place in St Thomas's since the fifteenth century, when the monks of Osney were the main producers. From 1434 onwards the University exerted quite an influence, even laying down a rota which specified the days on which particular brewers could brew. And the university colleges themselves used to produce beer, the most famous brewhouse being that at Queen's College, last used in 1939, when a brewer from Morrell's tended the last brew of 1100-gravity Chancellor Ale.

At the beginning of the twentieth century there were still five breweries left in Oxford (Hall's, now part of Allied Breweries, whose distribution depot stands close to Morrell's, mopped up the others). Morrell's was run by trustees between 1863 and 1943, but the family involvement remained strong (so strong, in fact, that a number of would-be buyers were warned off in the 1960s) and the present company is still virtually entirely family-owned. Today Colonel Bill Morrell is the chairman – the fifth member of the family to head the company – and three members of the sixth generation joined the board in 1983.

The tied estate has been painstakingly built up and now numbers 137, more than fifty of them within the Oxford ring road and every one within thirty miles of the brewery. Six have been built since the 1960s, and the Westgate in Oxford (the only managed house) was opened in 1982 to mark the bicentenary of Morrell's Brewery. Free trade has slowly been increasing, mainly within Morrell's trading area but with an excursion into the London area too.

Visitors to Morrell's country are likely to be confronted in some pubs (such as the Black Boy in the Oxford suburb of Headington) by an extraordinary range of cask-conditioned beers, all with their own individual characteristics. The most commonly available, perhaps, are Light Ale (1032), an excellent, pale, surprisingly sharp and thirst-quenching beer; Oxford Bitter (1036), pleasant and well-balanced; and Varsity (1041), a malty and full-flavoured premium brew. A lesser known member of the family is Dark Mild (1033), which is the Light Mild with added caramel, sold in only a few pubs. Two much stronger beers are Celebration (1066), a rich pale beer originally produced to mark Princess Anne's wedding and still to be found on draught in about a dozen pubs, and College Ale (1073), a very distinctive and powerful malty, fairly sweet pale barley wine, which is in some twenty-five pubs, mainly but not exclusively in the winter months. The bicentenary celebrations even saw a seventh traditional draught beer, 200 Ale, but this is no longer available. Nine-tenths of Morrell's pubs serve the beer traditionally, by handpump or direct from the cask, though some use blanket pressure or a cask breather to protect slow-selling beer.

The brewery itself, approached through an impressive wrought-iron arch guarded by gilded lions, has seen heavy investment in the past, not least with the construction of a 25,000-barrel-a-year lager brewery in 1986, designed to allow Morrell's to produce Harp Extra under licence, and envisaged as enabling overall production to stay at a level which will guarantee the future of Morrell's cask-conditioned beers. The ale brewery, its capacity double that of the lager plant, has its share of venerable vessels, including a 1901 mash tun on the ground floor, and the first hopback to be converted into a whirlpool, but the open cooler began to leak and had to be replaced. The beers are collected and pitched in traditional open fermenters, then 'dropped' into conical fermenters where, unusually, the ale yeast bottom-ferments. Outside the brewery, the waterwheel which had harnessed the waters of the Wareham Stream to provide power but which had been disused for some twenty years, was restored in 1982 and is now working again, a token, perhaps, of the company's confidence in its independent future.

PALMER'S
J. C. & R. H. Palmer Ltd, The Old Brewery, West Bay Road, Bridport, Dorset.
Pubs: 68. Beers: Bridport Bitter (1030), IPA (1039).

Palmer's are such a closely controlled family brewery company that they make other so-called 'family' companies appear positively overrun with outsiders. There are a mere four shareholders – A. A. J. Palmer, his wife and their two sons, who also constitute the board of directors. Yet the Palmer dynasty at the Old Brewery is of comparatively recent vintage, the brothers J. C. and R. H. Palmer having bought the business from Job Legg in 1896. The brewery was run as a partnership until 1975, when the present company was formed with J. C. Palmer's grandson as chairman. The history of the brewery, which stands below the confluence of the Brit and Asker rivers, has, however, been traced back as far as 1794, when S. Gundry began brewing in a converted mill on the present site.

Palmer's have never taken over another brewery company and are quick to emphasize that they themselves do not intend to be taken over (Tony Palmer is also vice-chairman of Eldridge Pope and there had long been fears that his retirement might presage a merger), stating that, 'Our aim is to preserve the business for our own children and theirs.' There is, however, a trading agreement, with Palmer's taking Faust lagers from Eldridge Pope in return for Palmer's IPA on a barrel-for-barrel basis.

The Old Brewery is one of the most picturesque in the country, with the bottling hall still operating under a thatched roof and a working waterwheel at the rear. The undershot waterwheel, dating from 1879 and forged at Helyear's foundry in Bridport, is curious in that the curved paddles do not fit the lugs forged into the wheel flanges – possibly because the height of the river weir was increased and the paddles had to be repositioned. The former maltings adjacent to the brewery is now a warehouse, and the mineral water factory is still in production, bottling a remarkably wide range of soft drinks.

Palmer's beers, widely noted for their refreshing and clean-tasting characteristics, are increasingly available in traditional form, with three-quarters of the sixty-eight tied houses having handpumps. Now only a few very low-turnover rural pubs are likely to dispense the beer by top pressure. The ales are Bridport Bitter (1030), often known as 'boy's bitter' because of its very low gravity but nonetheless a pleasant hoppy drink acccounting for over half of total sales, and the excellent IPA (1039), a rather darker and fuller-flavoured dry-tasting best bitter. In addition Tally Ho (1046), primarily a bottled beer but occasionally available

The Old Brewery, Bridport, Dorset

on draught for the summer tourist trade, is a dark and malty strong bitter sometimes dispensed direct from casks on the bar.

In 1945 there were about a hundred tied houses, but about a third have been closed. Most of these have been smaller country pubs where tenants have found it impossible to make a living. The most recent and most notorious closure was that of the Brandon Hotel in the commuter and retirement village of Netherbury, which Palmer's intend to replace with seven new private houses. Bridport, too, has suffered, and now there are only five pubs in the half-mile stretch between the brewery and the town hall, where once there were no fewer than fifteen. However, Palmer's are still well represented in Bridport, with seventeen pubs, and there are also six in Lyme Regis, five in Beaminster and three in Axminster, together with a scattering of country houses. Only one of the sixty-eight pubs is further than twenty-five miles from the brewery: the Ferry Inn in Salcombe, South Devon, in the centre of Palmer's main free trade area.

ST AUSTELL

St Austell Brewery Co. Ltd, The Brewery, St Austell, Cornwall. Pubs: 130. Free trade in Cornwall. Beers: BB (1031), XXXX Mild (1034), Tinners Ale (1038), Hicks Special (1050).

Hicks Special and Tinners Ale are hardly the best-known beers in Britain – indeed, it is likely that only those who have spent time in Cornwall recently will recognize the names, yet they signify a welcome resurgence in traditional brewing by a company which, whilst it is equally little-known, is still controlled by the descendants of the founder and is a significant owner of delightful Cornish pubs, from Bude to Penzance (with an outlier in the Isles of Scilly).

The business has its roots in Walter Hicks' decision, in 1851, to mortgage his farm at Menadhu and become a wine and spirits merchant in Church Street, St Austell. It was not until 1860 that he turned to brewing, and only in 1867, when he bought the London Inn, with its brewhouse and cellars, did he begin brewing on a large scale. Walter Hicks junior took over in 1890, commissioning the present brewery in Trevarthian Road, on the hill overlooking the town and St Austell Bay. Although there has been considerable modernization, the original brewhouse, completed in 1893, is still at the heart of the brewery. By 1893 fifteen

SOUTHERN AND SOUTH-WEST ENGLAND

The auction notice for the Sun Inn at St Austell – bought by Walter Hicks & Company in 1909 tied houses had been acquired, and twenty years later the number had increased fourfold.

Walter Hicks & Company was incorporated in 1910, with the shares being divided between Walter senior and his two sons and eight daughters. Today the shares are still controlled by descendants, though they are widely dispersed within the family, and the directors comprise the only surviving grandson of the founder and three fourth-generation representatives from different branches of the family. The brewery grew by acquisition after the Depression of the 1930s, merging with Christopher Ellis & Son of Hayle in 1934 (the Steam Brewery in Hayle was closed down and the present company name adopted) and acquiring six pubs from the Treluswell Brewery in Penrhyn when this ceased brewing and was taken over in conjunction with the Redruth Brewery in 1943. Since then piecemeal pub acquisitions have been balanced by losses, as in the case of Plymouth, where the former tied houses have been lost.

In 1930 St Austell produced six draught beers, namely XX, XXX, XXXX, PA, BB and IPA; now there are four, including two which derive their identity from the above list. XXXX (1034) is a fruity dark mild which is fairly popular in some parts of Cornwall but which is in only a few of the pubs and might not be viable were it not also bottled as brown ale; BB (1031), popularly known as 'boy's bitter', is a thin, light-flavoured bitter. For some time these were the only traditional beers, the stronger (1037) bitter, Extra, being produced only as a keg beer. The newest of St Austell's beers is Tinner's Ale (1038), an excellent dry-hopped bitter which is well balanced and at last provides a medium-strength bitter in traditional form. At the top of the range is Hicks Special (1050), introduced in 1975 in response to repeated requests for a strong draught ale – despite the failure of its predecessor, ESA (1060), first brewed for the Christmas trade in 1948 but discontinued on draught in the following year. Hicks Special is a mellow, warming, heavy bitter of real distinction, dry-hopped to retain a bitter aroma, and almost worth the journey to Cornwall on its own!

WADWORTH'S
Wadworth & Co Ltd, Northgate Brewery, Northgate Street, Devizes, Wiltshire.
Pubs: 146. Extensive free trade in southern and south-west

England. Beers: Devizes Bitter (1031), IPA (1035), 6X (1041), Farmers Glory (1046), Old Timer (1053).

Wadworth's six-storey redbrick tower brewery dominates one entrance to the market-place in Devizes, which is noted more for its restrained Georgian architecture. Yet the brewery is no more out of place in this celebrated townscape than are the company and its products in the affections of Wiltshire beer-drinkers, for Wadworth's is an outstanding example of a traditional country brewery with a fine regard for its brewing heritage and an outstanding range of quality real ales.

The first records of the Northgate Brewery date back to 1768, when Rose and Tylee were brewing at premises in Northgate Street which later became a furniture store. Various changes of ownership followed until in 1865 three coal merchants, George, John and William Sainsbury, took over; finally Henry Wadworth, a farmer and eccentric, bought out the Sainsbury brothers in 1875. Ten years later Wadworth was joined in partnership by his brother-in-law, John Smith Bartholomew, who had previously been head brewer at the Wallingford Brewery. 1885 also saw the construction of the present brewery on its imposing corner site adjacent to the market-place.

The firm was incorporated in 1889, the year which saw the start of modest expansion by take-over, virtually all of which was to be small-scale and very local. The result is that, even today, Devizes is dominated to an almost unhealthy degree by Wadworth's pubs. The first local firm to be swallowed was Humby's Southbroom Brewery in Devizes; then came the Bromham Brewery in 1896 and the Estcourt Brewery, Devizes in 1903. This latter brewery, owned by George Wild & Co., was offered for sale with its six pubs in 1902, but Wadworth's bought only the brewery itself and the adjacent Bell Inn.

In 1919 the Three Crowns Inn in Maryport Street, a home-brew pub run by J. W. Phipp, was purchased, further strengthening Wadworth's hold on Devizes itself and also offering an example of the process which led to the virtual extinction of pubs brewing their own beer on the premises. Ten years later Henry Wadworth, the founder, died; he had been chairman of the company for the forty years since its inception in 1889. He was succeeded by John Bartholomew, son of his brother-in-law. There was no change in policy, however, and the company continued to expand with occasional modest acquisitions.

The most recent of these acquisitions was that of Richard Garne & Sons, whose picturesque though rather cramped fifteenth-century brewery in Sheep Street, Burford, is still standing and indeed is still used as a distribution depot. Garnes was a long-established Oxfordshire brewer, dating from at least 1798, with nine pubs in and around Burford. Its purchase in April 1969 was an uncharacteristic move from Wadworth's, further away from their Devizes heartland than any previous acquisition, and with only the nine pubs to supply (the Sheep Street brewery was soon shut down) it cannot have made immediate economic sense. Yet in the longer term it has given Wadworth's a presence in an important free-trade area and contributed to the present position, where free trade accounts for two-thirds of turnover.

An adherence to traditional methods has always been a feature of Wadworth's operations. Until the 1960s, when it became uneconomic to continue, locally grown barley was malted in the brewery's own maltings in Northgate Street. Local deliveries have always (except for a break of five years in the 1960s) been made by horse-drawn dray, and Wadworths' teams of shirehorses are a particular attraction in the streets of Devizes. The brewery is amongst the very few which retain a cooper's shop where the wooden casks into which much of the brewery's output is racked are repaired and even rebuilt.

A tied estate of 146 pubs has been built up, heavily concentrated in central Wiltshire but with outlying concentrations in Bristol (where several pubs have recently been acquired), Cheltenham, the Burford area and Andover. Many of the pubs are delightfully situated in rural areas and remote villages, and many are highly traditional, even to the extent of serving beer direct from the casks – as, for example, at the Raven Inn at Poulshot, west of Devizes.

At all except one of the pubs at least one of the remarkably wide range of traditional draught beers is available. The beers comprise Devizes Bitter (1031), admittedly weak but a well-hopped beer which is ideal for lunchtime or 'session' drinking; IPA (1035), seriously under-rated by many but a marvellously light bitter with a delicate flavour; 6X (1041), one of the classic real ales of England and promoted as such, full-bodied and quite sweet, and accounting for seventy per cent of production; Farmers Glory (1046), introduced in 1984 and immediately successful, much darker than the other beers but with a fifty per cent higher hopping rate and hence still comparatively dry-tasting; and Old Timer (1053), a gloriously rich and fruity pale strong ale which is more generally available in the winter months.

6 Wales

Drink the remaining ales of Wales while you can. Following the take-over of Border Breweries by Marston's in 1984, only four 'independent' firms survive, and of these Felinfoel are 49.5 per cent owned by their Llanelli rivals Buckley's, who in turn have two of their brewing brothers, Whitbread and Belhaven, as major shareholders. Crown Brewery, one of two surviving clubs' breweries, and Brain's complete the quartet. Even the Big Six brew very little beer in the Principality, with Welsh Brewers (Bass) in Cardiff providing the only locally brewed real beer; Allied in Wrexham and Whitbread at Magor produce only processed beers and lager.

How different things were a quarter of a century before Border's demise. Some twenty brewers competed in South Wales alone, of which Evan Evans Bevan, Rhymney Breweries and the Ely Brewery all fell to Whitbread, and Hancock's of Cardiff and Swansea, Webb's (Aberbeeg) and David Roberts of Aberystwyth all eventually became a part of Bass. Though these two Big Six brewers consequently dominate many Welsh drinking areas (notably the former mining valleys), there are welcome pockets of excellence in Cardiff (Brain's) and Llanelli, and wide areas where free houses provide an excellent choice of beers – eastern Gwent and the Brecon area, for example.

In mid-Wales Banks's and Wem ales are widely available, while further north several north-western breweries have acquired sizeable tied estates. Burtonwood Brewery, Lees and Robinson's all come into this category. New independent brewers have fared badly in Wales, with Gwynedd Brewery, Cestrian (Clywd), Powys Brewery, Red Kite (Aberystwyth) and the Afan Brewery (Port Talbot) all going to the wall, but some excellent beer is brewed by the few who have fared better, such as Silverthorne's in Gwent and Samuel Powell in Newtown (a long-established company which ceased brewing in 1956 and became wholesalers but began brewing again on a different site in the 1980s).

BRAIN'S
S. A. Brain & Co Ltd, The Old Brewery, St Mary Street, Cardiff, South Glamorgan.
Pubs: 120, heavily concentrated in Cardiff. Free trade in south-east Wales and Avon. Beers: Red Dragon Dark (1035), Bitter (1035), SA (1042).

Brain's has a cult following in Cardiff, almost rivalling the National Stadium as the focal point of local affection. Yet in the adjacent South Wales valleys the beers have only recently gained any acceptance at all, and further afield the impact of the three excellent beers is even more limited compared with the more renowned brews from Fuller's, Greene King, Ruddle's and others.

The first known brewery buildings on the St Mary Street site, right in the centre of Cardiff, were producing beer by 1713, and the brewery has been in continuous use since then, although the last of the original buildings was pulled down at the end of the First World War. The first brewery probably supplied the Albert Hotel, behind which it stood, and one or two adjacent pubs. There were frequent changes of ownership until, in 1882, John Griffen Thomas was bought out by his brother-in-law, Samuel Arthur Brain, together with his uncle, Joseph Benjamin Brain.

S. A. Brain, who had previously managed the Phoenix Brewery in nearby Working Street, became chairman of the company when it was registered in 1897, but in the fifteen years before that he had been largely responsible for a period of rapid expansion. In 1882 only eleven public houses were supplied with beer from the Old Brewery, and only a hundred barrels a week were brewed. Brain quickly recognized the limitations of the Old Brewery buildings and replaced them in 1887, at a cost of £50,000, with what was then the largest brewery in South Wales. The risk was amply justified, for by 1897 Brain's controlled seventy-four pubs, and output had increased tenfold, to over a thousand barrels a week.

The 1887 brewery is still in use today, though it has been surrounded by later development; indeed, in 1979 the brewing and fermenting capacity was increased by half as a result of further intensification of use of the already very cramped site. In addition another bold move in the inter-war period saw the construction of the New Brewery in suburban Cardiff, purely to produce bottled beers (these, especially two-pint flagons, are still

The Old Brewery buildings of 1713 (demolished in 1919)

curiously popular in Cardiff). In 1965 the New Brewery also began to brew a low-gravity keg beer, and since 1985 Faust lager has been brewed there.

Only two of the original eleven public houses are still open for trading: these are the Albert Hotel, a much-renovated structure adjacent to the Old Brewery, and the Golden Cross, a listed building with a glazed-tiled façade and superb tiled murals inside. The Golden Cross was recently threatened with demolition, but spirited opposition led rather more by CAMRA than by the brewery led to its reprieve. The remainder of the original pubs have succumbed to demolition through redevelopment schemes, including the legendary Taff Vale in Queen Street.

Three traditional draught beers are produced at the Old Brewery, and when they are kept well they are quite superb, as well as being notably good value for money. Brain's Dark, occasionally still known as Red Dragon (1035), is a subtle, full-flavoured malty beer which despite its colour is almost too dry-tasting to qualify as a mild; bitter (1035), usually referred to as 'light', is a well-hopped, truly bitter and refreshing drink; SA (1042), named after the founder, is a magnificent premium bitter, full of flavour and still quite hoppy in character – though Brain's are perhaps going too far in stating that, 'New drinkers are advised to treat it with respect' – it isn't *that* strong. Finally, the Crown Inn at Skewen near Neath is the only pub to sell a brewery mix of SA and Dark, known as MA.

BUCKLEY'S
Buckley's Brewery plc, Gilbert Road, Llanelli, Dyfed.
Pubs: 180. Beers: XXXX Mild (1032), XD Mild (1032), Best Bitter (1036).

Colonel Kemmis Buckley, chairman of Buckley's until 1983, has attributed the company's survival to its location, 'far enough to the west in Wales to have deterred large predator brewing companies in time past from making bids for us'. Undoubtedly this peripheral location has had the effect of keeping Buckley's from the centre of the brewing stage: few realize that it is a comparatively large independent brewer, with 180 pubs (most serving handpumped beer), a widespread free trade and a fierce pride in the beers which are produced.

The Buckley connection with Llanelli dates back to 1794, when the Reverend James Buckley, a disciple of John Wesley, visited Henry Child, philanthropist, brewer and maltster (the owner of the Old Mill maltings, newly constructed west of the brewhouse yard). Child had recently bought public houses and a maltings in Llanelli and in 1799 built the first brewery on the present site, close to a well which supplied brewing-water until the 1920s (now the town water supply is used). The Reverend James married Child's daughter Maria, and when Henry Child died in 1824 his son-in-law combined the management of the brewery with his vocation as a Methodist minister – a fairly eccentric combination!

A public company, Buckley Brothers Ltd, was formed in 1894 by James Buckley's grandsons, Joshua and Joseph, and it immediately began to expand, acquiring Bytheway's Brewery on Emma Street, Llanelli in 1896 and the Carmarthen United Brewery when it ran into financial difficulties four years later. This latter acquisition extended the spread of the company's outlets as far as Merthyr Tydfil, but the outlying pubs were quickly sold to maintain a reasonably compact trading area. Buckley's have bought only one other business since then (the Llan mineral-water factory in nearby Llangennech), though this is not for want of trying: in 1965 they acquired 49.5 per cent of the shares in Felinfoel Brewery, but their bid to take over the only other surviving independent brewer in West Wales was rejected. Buckley's have retained their shareholding, however, and treat Felinfoel as an associated company.

Buckley's themselves have intermittently been the subject of take-over speculation. They were amongst the first to seek shelter under Whitbread's notorious 'umbrella', and today the big brewer has a seventeen per cent share stake – almost double that of the Buckley family. Financial institutions hold a massive stake (the six largest between them control twenty-nine per cent), and a newcomer to the list is Nazmu Virani, chairman of Belhaven Brewery, who acquired a 6.2 per cent stake in 1985, prompting a period of uncertainty but, as yet, no further developments.

Major developments have taken place at the brewery in recent years, with the maltings closed in 1961, metal casks replacing their wooden equivalents in the mid-1960s, the brewery completely remodelled in the 1970s and the brewhouse re-equipped, with the most recent change being the replacement of an ageing hopback with a whirlpool in 1984. Not everything has changed, however: the strain of yeast is over fifty years old and is used by a number of other breweries, including nearby Felinfoel and, at times, Wadworth's and Everard's.

In 1890 two of the most popular Buckley products were BA pale ale, reputedly a favourite drink of Welsh miners, and Pale Bitter Ale, 'a bright, sparkling and well-flavoured beverage, very popular at Tenby and other seaside resorts', according to Alfred Barnard. There were, of course, other stronger beers, including an Old Welsh Ale which was matured in carriage casks. As recently as 1983 there were four draught beers from Buckley's,

The East Yard at the Llanelli Brewery in 1890

including Gold (1042), a pale and full-bodied cask version of a keg bitter which made only a fleeting appearance in traditional draught form. Now there are only two (the milds being virtually identical) since Standard Bitter, a very thin brew, was axed in 1985. The survivors are Best Bitter (1036), a well-balanced but not especially memorable beer, and XXXX Mild (1032), dark and rather sweet. An extra-dark version of this, known as XD, is made available in the Swansea area. The mild is available in certain Whitbread pubs in South Wales, as far east as Cardiff, and is also sold in Ansell's houses in the area as Ansell's Dark.

CROWN BREWERY
Crown Brewery Co Ltd, Pontyclun, Mid Glamorgan.
Pubs: 5. Massive free trade, overwhelmingly in clubs. Beers: Black Prince (1036), SBB (1036), 'House' bitter for individual pubs (1041).

Crown Brewery, though it is now building up its own tied estate and increasing its presence in free houses, is still essentially a clubs' brewery – one of only two left in the country, and the only one to brew traditional draught beer – whose primary aim is to provide reliable supplies of low-priced beer to clubs both in South and West Wales and also further afield, in the London, Bristol, Southampton and Birmingham areas.

The South Wales and Monmouthshire United Clubs Brewery Company Limited, as it was then somewhat ponderously known, was incorporated in 1919 and in that year purchased the Crown Brewery at Pontyclun, previously run by D. & T. Jenkins Ltd. The reason for the company's formation was dissatisfaction in the clubs of South Wales with the high prices charged by brewers, which were regarded by the local Working Men's Club and Institute Union as 'entirely unwarranted and unnecessary', and erratic supplies of beer to clubs during and immediately after the Great War. Shares in the brewery were (and still are) available only to clubs and individuals who were members of the CIU.

The early 1920s saw many clubs breweries established for similar reasons, though most were short-lived affairs and only Crown Brewery and its larger northern counterpart, the Northern Clubs Federation Brewery, are still operating. Amongst the longer-lasting clubs' breweries were the Walsall & District, founded in 1920 and taken over by Charrington in 1960; the Northants & Leicester Co-operative, where brewing commenced in 1921 and continued until as late as 1969, and the Metropolitan & Home Counties, which bought the Anchor Brewery in Maidstone in 1920 but was supplied with beer by the Northants & Leicester Co-operative from 1946 and eventually went bankrupt in 1956.

Initially about 200 barrels a week were produced, but as clubs began to recognize the benefits of forging links with the new brewery – assured supplies, good-quality beer and low prices (partly in the form of a bonus for each barrel purchased) – demand began to grow rapidly. By 1936 bottled beers had been introduced and production was up to 500 barrels a week; new fermentation capacity was added as output soared to 900 barrels a week in 1938; and by the late 1940s the weekly figure was 1,200 barrels and the brewery was almost collapsing under the weight, tie-bars being added to keep the walls from collapsing. Urgent action was required, and the present brewery was completed in February 1954.

The original product was Clubs Pale Ale (CPA), which by the time it was finally withdrawn in 1984 was a run-of-the mill light mild with an original gravity of 1033. Special Best Bitter – SBB (1036) – was introduced in 1960 and is now the mainstay of traditional beer production; a malty, instantly recognizable brew at its best, it is also available in a dark form as Black Prince (1036), an unmemorable mild which replaced Crown's true dark mild, 4X, in 1985.

The brewery introduced its first keg beer in 1969, shortly after filtered brewery-conditioned beer had been produced, and some eighty per cent of output now consists of these processed beers, largely because the low barrelage of many clubs and the high turnover of club stewards makes them unwilling to accept cask beer. The present premium keg bitter is Great Western, at 1041 the same gravity as Crown's newest real ale, sold as Star Bitter at the Star Inn, Treoes, and as St Oswald's Bitter at the St Oswald's in Port Talbot. These two pubs, the first bought from Bass and the second a converted club, are amongst the five now owned by Crown; another, destined to become the brewery showpiece, is the Brunel Arms in Pontyclun, a former social club which went into liquidation owing Crown £60,000. Free trade in pubs and clubs is growing, with depots opened in Narberth, Dyfed and Charlton in south London, but the promise of more pubs to come, not only in Wales but also in the West Country, is even better news for supporters of SBB.

FELINFOEL
The Felinfoel Brewery Co Ltd, Felinfoel, Llanelli, Dyfed.
Pubs: 74. Extensive free trade, mostly in West Wales. Beers: XXXX Mild (1031), Bitter (1035), Double Dragon (1040).

An unusual brewery company, with virtually half its shares owned by its closest competitor and a stubborn reluctance to offer its beers in traditional form in its own tied houses, Felinfoel is nevertheless known as a producer of quality ales, notably the premium bitter Double Dragon. Its beers are well known elsewhere too, for Felinfoel have recently begun to export bottled beers, occasionally to Penang, Kuwait and Toronto and on a more regular basis to San Francisco, where the barley wine known as Hercules Ale – not sold in Britain – has been particularly well received. The brewery's trademark, a red dragon on a green background, is a familiar and welcome sight outside free houses in South and West Wales and even further afield, for in the free trade the beers are almost always served traditionally.

The story begins in a home-brew pub, the King's Head in Felinfoel, in the 1830s. This was where David John, a local tinplate manufacturer, turned to brewing with such success that he was soon supplying beer to a number of other pubs in Felinfoel and Llanelli. By 1878 demand had grown to such an extent that the present brewery was constructed, and at the same time Felinfoel began to buy pubs, gradually widening the trading area. Somewhat later, David John's grand-daughter married a John Lewis, and the Lewis and John families combined to run the brewery.

This amicable relationship was not to last, however. In 1965 there was a major disagreement between the two families, with the result that the John shares (990 out of a total of 2,000) were offered for sale. The major brewers, who at that time were buying scores of their smaller competitors, were immediately interested, but the shares were eventually acquired by Felinfoel's neighbours Buckley's, who made a bid for the remainder of the shares. The Lewis family remained firm, however, and they have since formed a holding company, Gustmain, to preclude any unwelcome take-over bid. Nevertheless, Buckley's still hold 49.5 per cent of the shares.

Felinfoel's major claim to fame is, perhaps, that they were the first British brewery to sell beers in cans, in 1935. The primary objective was to assist the Llanelli tinplate industry, which had been hard hit by the recession, though this was a practical rather than an altruistic move, since the brewery directors were also directors of a tinplate mill. Demand was limited until the Second World War, when the NAAFI took the beer and shipped it around the world, but since 1945 it had reverted to a relatively low level until recent years, when multiple stores and off-licence chains have begun to stock Double Dragon. The beer was canned in Brasso-type tins until 1948.

Felinfoel's beers – almost universally available on handpump in the free trade but unpressurized in only a quarter of the seventy-five tied pubs, which are quite widely spread throughout Dyfed – are XXXX Mild (1031), a dark brew of no great distinction which accounts for only five per cent of production, a pleasant light bitter (1035), very popular in the Llanelli area, and Double Dragon (1040), an excellent full-flavoured ale which accounts for almost half of total production and is the beer which is demanded almost exclusively by the free trade. Double Dragon won the

A Felinfoel bitter label

award for the best cask-conditioned beer at Brewex in 1976; to celebrate, Felinfoel renamed a renovated pub in Morriston, Swansea, the 'Champion Brewer' (though it sells no traditional beer!).

The brewery has been extensively modernized in the last decade, with equipment that had survived since 1878 at last being replaced. The greatest advance was the replacement of the original open, coal-fired copper, which used to pollute the whole brewery, with a modern stainless steel version. It is all a far cry from the brewery in the 1950s, when the beer had a poor reputation and the company, some of whose pubs were selling only eighteen gallons of beer a week, was on the verge of bankruptcy. Now the brewery has been updated and the beers have a high reputation; one can only hope that the promised renovation of the pubs will include the provision of more traditional beer.

7 The West Midlands

No overall pattern is discernible in West Midland brewing. Birmingham drinkers are notoriously badly served, only Davenport's of the six brewers who were still producing beer in 1960 battling on into the 1980s as an independent force (and then attracting the attentions of first Wolverhampton & Dudley Breweries and then, fatally, Greenall Whitley). Mitchell's & Butler's (a Bass subsidiary) produces rather bland beers, and Ansell's Aston brewery, once the home of one of the best milds in the country, produces none at all, having been shut down in 1981 after a prolonged strike.

The Black Country fares much better, Wolverhampton & Dudley brewing good, cheap ales at its Banks's and Hanson's plants, two tiny firms (Batham's and Holden's) brewing some outstanding beers, and the Old Swan in Netherton still famous for its home-brew. Simpkiss, however, gave up the ghost in 1985, allowing itself to be taken over and closed by Greenall Whitley. Greenall's Shropshire subsidiary, Wem, is a substantial producer of traditional beers, and Shropshire also has two firmly established home-brew houses, the All Nations in Madeley and Three Tuns in Bishops Castle. To the south is the fine country brewery at Hook Norton, near Banbury, and in the east, in the brewing capital of Burton-on-Trent, regional brewers Marston's are one of four surviving traditional breweries (the remnants of the twenty-six existing in 1900). Bass and Allied Breweries both have major breweries here, Allied's Ind Coope brewery in particular producing a string of similar ales for marketing companies in the South and Midlands. Burton's fourth brewery is now being promoted as a 'heritage' brewery and produces contract ales for its former owners, Everard's, whose pubs are all in the East Midlands. Relatively few new brewers have set up except in the south-west of the region.

Not only Ansell's were famous for their mild, for together with north-west England and the cities of South Wales this is one of the last bastions of the celebrated but threatened beer style. Highgate Brewery in Walsall, now owned by Bass, has the distinction of being the only brewery in Britain to produce mild ale alone, and its excellent dark mild is very popular. Banks's and Hanson's, indeed, produce much more mild than bitter (and not just for a few pubs: their tied estate numbers some 800), and the quantity of mild brewed is still increasing.

ALL NATIONS
All Nations Inn, Coalport Road, Telford, Madeley, Shropshire.
Home-brew pub. Beer: All Nations Home Brewed (1032).

Dating from 1789, when it was built by Christopher Bagley, and in the hands of the Lewis family for over fifty years, the All Nations is a home-brew pub which has survived perhaps more because of its out-of-the-way location, standing back from and above a minor road leading down from Madeley to the Severn Gorge, than as a result of the nevertheless admirable persistence of its proprietors. What survives now is neither picturesque nor particularly geared to the tourist market, but the All Nations has generated lasting local affection, with a fiercely loyal band of regular customers, and it is as much a part of the area's industrial history as the much-vaunted Blists Hill open-air museum across the road.

The pub has had to survive two major upheavals this century, the first in 1934 when W. H. Lewis acquired it from the founder's descendants, the second much more recently, when Bill Lewis's death in 1975 created uncertainty as to the future of the home brew. Fortunately Bill's widow Eliza, who had actually carried out the brewing for the previous forty years (and had helped her father at the nearby Bird in Hand in Ironbridge before that), was determined to carry on, and with the help of her son and daughter-in-law the once-weekly brew has been saved. Mrs Lewis is still the licensee of the pub, although the brewing is now in the hands of her son Keith.

The brewery, comprising mash tun, copper, cooler and fermenting vessels, lies immediately behind the pub, and it was here that Mrs Lewis used to transfer the brew, all seven barrels (250 gallons) of it, from cooling to fermenting vessel using a hand ladle. Now the hard labour has been replaced by an electric pump, and a much more recent improvement has involved repairs to the fermenters – which meant a break in production in the summer of 1985 (Holden's beers replaced the home-brew, but happily this is now very much back in production).

As for the beer itself, it clearly commands a sizeable local following, but alas many visitors will be surprised to find it rather thin and lacking in real character. It is a surprise to find the Good Beer Guide describing it as a 'pale ale'; pale it certainly is, even lighter than Boddington's bitter in that beer's heyday, but in terms of taste it is a classic light mild, rather sweet and malty and only very delicately hopped. The pump clip in the plain, homely bar describes it simply as 'All Nations Home Brewed', neatly avoiding problems of classification. One very definite attraction is the price, which is spectacularly low (57p early in 1986, 5p less than draught mild from Joseph Holt's, a brewery widely known for its exceptionally low prices).

BANKS'S AND HANSON'S

The Wolverhampton & Dudley Breweries plc, Park Brewery, Lovatt Street, Wolverhampton (Banks's). Also trade as *Julia Hanson & Sons Ltd* (Hanson's).

Pubs: 800. Beers: Banks's Mild (1035), Hanson's Mild (1035), Hanson's Black Country Bitter (1035), Banks's Bitter (1038), Hanson's Bitter (1038).

In most respects Wolverhampton & Dudley – usually referred to as Banks's – are, despite their size, a superb example of a traditional brewery company, with excellent real draught beers, traditionally brewed in old-fashioned breweries, and very low prices in the pubs. Banks's brew no lager, buying in their supplies from Harp (they were formerly members of the Harp consortium), and astonishingly they brew nearly three times as much mild as bitter. But in just one respect there are niggling doubts. This relates to the company's quite explicit ambition to grow bigger – W & D 'makes no secret of its desire to spread further' and has 'an active policy of acquiring sites and businesses throughout its trading area'. Hence the bid for Davenport's of Birmingham, rejected in 1983 but renewed, again unsuccessfully, in 1986.

The company effectively started by amalgamation in 1890, when three Black Country enterprises, Banks & Company (who already occupied the Park Brewery site), George Thompson & Sons of the Dudley and Victoria breweries, and C. C. Smith of the Fox Brewery in Wolverhampton, joined forces under the present name. The Banks family had become maltsters in 1840 and brewers rather later, but they built the Park Brewery in 1875 and advertised as follows: 'Banks & Co, Brewers, Chapel Ash, Wolverhampton, beg to inform their customers friends and the public, that they have removed from Newbridge to their new Brewery at Chapel Ash, and are now in a position to supply their October brewed ales, which they can strongly recommend. They apologise for the occasional delay in delivery caused through the insufficiency of their late premises, but now having a much larger plant trust the inconvenience will not occur again.' The Thompson family spearheaded the early growth of the company, Edwin Thompson becoming managing director in 1894, and his great-grandson David Thompson maintained the tradition when he was appointed to the same post in 1986.

Early in the twentieth century the company gradually expanded, first locally and then further afield, away from its Black Country base. The North Worcestershire Breweries, operating from Stourbridge and Brierley Hill and owning 135 pubs, were bought in 1909, and John Rolinson & Sons of Netherton followed in 1912. 1914 saw the acquisition of the Kidderminster Brewery Company, with 126 pubs in a new area for the company, 1917 the purchase of the City Brewery, Lichfield and its 200 pubs (Banks's still operate a maltings here), and 1928 the acquisition of Robert Allen of Worcester. Finally, in 1943 Julia Hanson & Sons of Dudley was acquired; the Hansons, who had originally been

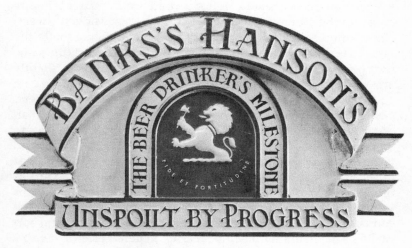

The Banks's and Hanson's logo

wine merchants, had entered brewing in 1889 when the Peacock Hotel and the brewery behind it were purchased.

More recently the company has concentrated on building up its tied estate, often by building brand-new pubs. This trend has been accelerating: in 1984 alone six houses were built, eight individually acquired and in two more substantial deals W & D also took over Cheshire Inns, with fourteen pubs in Cheshire and the Wirral, and bought five houses in the Manchester area from Wilson's Brewery. And following this impressive boost to the tied estate, there were another seventeen new houses opened in 1985 and thirteen in 1986. Although two-thirds of Banks's and Hanson's beer is sold within twenty miles of the brewery, the trading area now stretches from Manchester to Bristol, and from Leicester to Aberystwyth.

Despite all these activities the company remains committed to the production of traditional beer: not for nothing has the advertising slogan 'unspoilt by progress' been adopted. The company still owns two maltings in order to control the quality of malt supplied, and although the two breweries have been virtually rebuilt since the 1960s traditional techniques and vessels have been retained. The mash tuns, dating from 1963 and 1980, are made of copper, and much of the beer is still fermented in circular oak and deal fermenting vessels in what is known as the Round Room.

The draught beers are dry hopped; hence the superb hoppy aroma of the bitter and the lack of cloying West Midland sweetness with the mild. The recipes are the same at both the breweries, but differences in brewing liquor and yeast (the same sixty-year-old strain is used in both breweries but it is slightly affected by its environment) give marginal variations in palate between Dudley-brewed and Wolverhampton-brewed beers. Mild (1035) is a medium-dark brew, dark reddish rather than black, which is exceptionally popular locally and has a nutty, mellow flavour. The bitter (1038) is a dry and crisp drink. And Hanson's Black Country Bitter (1035), launched in 1985, though it upset Holden's because of the use of the Black Country name, has brought an extra beer to many Banks's and Hanson's pubs (despite its name, it is actually brewed at Banks's). All the company's pubs have temperature-controlled cellars, and virtually all of them dispense the beer by metered electric pumps, which the company feel are more hygienic than handpumps (and which also dispense an exact pint, with the bonus of a creamy head).

BATHAM'S
Daniel Batham & Son Ltd, Delph Brewery, Brierley Hill, West Midlands.
Pubs: 8. Beers: Mild (1036), Bitter (1043), Delph Strong Ale (1054).

Its brewery half-hidden by one of the Black Country's most famous drinking-institutions, the Vine (much better known as the Bull and Bladder), Batham's is paradoxically both bigger than it looks and smaller than it was. With only eight pubs the firm is a minnow of the brewing industry, but its magnificent bitter is revered over a wide area despite (until recently) only minimal attempts to promote it.

The Delph Brewery was founded in 1876, but it was not until 1905 that Batham's occupied it. The original Delph firm became part of the Worcestershire Brewing & Malting Company in 1896, and the brewery became surplus to requirements. Batham's began in Cradley Heath, where Charlotte Batham was a shopkeeper and home-brewer of repute; her son Daniel, founder of the firm, began with a home-brew pub in Netherton in 1877 and later took a pub in his native Cradley Heath before moving to Brierley Hill. At one time there were nineteen Batham's pubs, but these have gradually been whittled away as death duties have taken their toll (most recently in 1974), and now only eight remain.

There is, however, no further contraction in sight, for Arthur Batham, who after his father (Daniel's son) died in 1974 'felt like giving up there and then', is now more determined to keep going, not least because his son Tim has in turn taken over the brewing. The firm has even begun slowly to expand again – free trade has been obtained, a new beer added and output increased. Despite criticism (unjustified) of the beer's flavour, and concern (justified) over the intentions of other companies, Batham's future looks markedly more secure.

Batham's superb bitter (1043) is the most memorable of their beers, and it has always been one to provoke strong feelings. Recently described as 'sugar water', to the irritation of Black Country drinkers weaned on sweetish beers, it is not easy to describe. Certainly it is fairly sweet, but it also has a crisp, bitter tang and is a delicious, strong, very pale and well-balanced beer.

A beer mat advertising Batham's magnificent ales

Sparkling pale ale, one of Batham's range of bottled beers, discontinued in the 1950s

The mild (1036) is deceptively strong for its type and makes a fine, dark and smooth alternative to the bitter – though it is not especially popular locally and could in future be threatened by falling sales. The third beer, revived in the winter of 1980–81, is Delph Strong Ale (1054), a powerful winter ale based on an old brew which was discontinued many years ago. There are no bottled beers (and have been none since the mid-1950s).

The eight pubs cover a surprisingly wide range, from basic town locals to fine country pubs (at Kinver, Chaddesley Corbett and Shenstone, near Kidderminster), although by far the most famous is the brewery tap, popularly referred to as the Bull and Bladder after a butcher's shop which used to form part of the pub's frontage. Across the top of the pub runs the Shakespearian quotation 'Blessing of your heart you brew good ale', and the adjacent brewery offices carry the inscription 'The birthplace of genuine beer'. Few would argue with such sentiments, at least not after close acquaintance with the bitter. And the beer is becoming available to a wider clientele, too, with a small but growing number of free houses stocking Batham's and a selection of local Ansell's pubs taking the bitter as part of a trading agreement.

DAVENPORT'S
Davenport's Brewery Ltd, Bath Row, Birmingham 15.
A subsidiary of *Greenall Whitley*, with 130 pubs. Beers: Mild (1035), Bitter (1039).

Davenport's has one of the most unusual of brewery company histories, and it is still probably best known for its 'Beer at Home' service, even though this was of declining significance for many years and was finally sold off in 1985. It also staved off take-over by Wolverhampton & Dudley in 1983 and 1986, only to find Greenall Whitley stepping in to rob Birmingham of its sole surviving independent brewery company, the only home-grown alternative to the blandishments of the M & B and Ansell's duopoly.

The firm can trace its origins back to 1829, when Robert Davenport was brewing in Brearley Street, Hockley, and also owned other licensed premises. The business expanded piecemeal, acquiring the Fox & Dog in Princep Street and several other pubs, but brewing was gradually concentrated at Bath Row, where Robert's son John had bought Bath House, a well-known mansion, in 1852. The gardens and cherry orchard behind the mansion have gradually been built over, but some of the original Bath House bedrooms are still intact and used as offices. Davenport's great leap into the Beer at Home business was begun in 1904, allegedly because Baron John Davenport disliked the sight of children being sent into pubs with a jug every time their parents wanted a drink at home.

Whatever the reason, delivery rounds began, and in order to finance the move, Davenport's sold most of their pubs. In its heyday, between the wars, it was a flourishing business operating from eighteen depots, from South Wales to Leeds and London to the Wirral, and taking four out of every five pints brewed at Bath Row. But increasing competition from off-licences and supermarkets meant that results were consistently disappointing from the 1970s onwards, and the business was sold in 1985, though Davenport's beer will still be supplied by the new owners.

Until 1961 Davenport's remained a large brewery with very few pubs, although a few 'flagship' houses had been bought close to their Beer at Home depots. But the purchase of Dare's Brewery, close by in Birmingham, brought forty pubs, mainly in and near the south of the city, into the fold, together with Dare's keg bitter, Drum Bitter, which became the biggest seller until 'the drum you can't beat' was eclipsed by traditional draught bitter in the late 1970s. An even bolder move followed the closure of Thornley Kelsey's Leamington brewery in 1968, for Davenport's snapped up twenty-nine of the sixty-eight pubs, mainly in the Warwick and Leamington areas. And in the 1970s the brewery began to build pubs in the Birmingham suburbs (where their houses had been few and far between because of the Ansell's and M & B stranglehold) and also began to buy pubs much further afield, especially in the Bristol area.

All this expansion attracted take-over speculation in 1980, when rumoured approaches from Grand Metropolitan and Greenall Whitley were strenuously denied, and again in 1983, when Davenport's future was much more seriously threatened as Wolverhampton & Dudley built up a nine per cent holding and then made what the board dismissed as a 'wholly unacceptable' bid. Although W & D pledged the future of the brewery (but not all of the beers), and although major institutional shareholders accepted the offer, the bid failed, largely because of the last-minute intervention of the Whitbread Investment Company. The price of independence was, therefore, a place under the 'umbrella', and when Whitbread's holding was later increased to eighteen per cent (compared with some twenty-seven per cent in the hands of the Baron Davenport's Charity Trust and other Davenports pension funds and charities, and 15.7 per cent owned by W & D), the Wolverhampton firm decided to bid again in 1986. Their offer was pitched at such an attractive level that Davenport's Board, though opposing the take-over attempt, could hardly hope to retain the firm's independence – and indeed a slightly higher bid from Greenall's was quickly accepted by the Board and their allies.

The present traditional beers comprise Mild (1035), which despite a recent two-degree rise in original gravity, said by the brewery to have made it 'stronger and creamier', is bland and fairly sweet, and bitter (1039), a really superb well-hopped brew with a noticeably dry, bitter taste, a gloriously hoppy aroma and a full flavour. The beers are now usually served traditionally – a welcome change from the practice a decade ago. Amongst the bottled beers special mention should be made of the two excellent strong ales, Top Brew (1071) and the even stronger Top Brew De Luxe (1074). A significant proportion of Davenport's trade lies in packaged beers, not only their own brews but also contract bottling and canning for an impressively large number of other brewers; the latest investment is £1 million in a high-speed line to fill large plastic bottles.

HOLDEN'S
Edwin Holden's Ltd, Hopden Brewery, George Street, Woodsetton, Dudley, West Midlands.
Pubs: 16. Beers: Black Country Mild (1036), Black Country Bitter (1039), Special Bitter (1052), Old Ale (1080).

Holden's, perhaps the least-known of the small and decreasing band of Black Country brewers, have no beers of national repute, or even pubs such as Batham's Bull & Bladder or the incomparable Old Swan Inn, both of which feature on offbeat tourist itineraries. What they do have is an increasing tied estate, mostly consisting of down-to-earth, genuine Black Country locals, and the widest range of beers in the area.

The most popular of the beers, even in the real-ale revival days of the late 1970s, was Holden's Golden, a keg bitter widely regarded as not much better than the appalling slogan which advertised it as 'the beer with the glow'. Now, however, it is Black Country Bitter (1039), a fairly sweet, malty and quite dark brew, which sells best of all, even though the Black Country is a mild-drinker's paradise. Holden's own dark mild (1036), full-bodied and decidedly sweetish, sells about half as much as the bitter. Perhaps the most welcome beer is Special Bitter (1052), a malty premium bitter first brewed in 1972 for a brewers' exhibition and now regularly available in the majority of the pubs. Old Ale (1080 at present, though it was 1075) is a splendid dark old ale which is normally available for only a limited period around Christmas. And December 1985 saw an even stronger ale, XL (1092), launched in the tied houses, though its future is uncertain.

The present managing director is Edwin Holden, the great-grandson of the founder. Together with his mother he owns the company – and a separate bottling company – and there were fears after his father's death in January 1981 that take-over and closure of the brewery were imminent. But as Edwin Holden says, 'We could have sold out in the 1960s but we held on. We feel we have a responsibility to guard the business for future years.' By 1981 he had already been working at the Hopden Brewery for fifteen years (having previously served his apprenticeship with

A Holden's beer label

Delivery by horse-drawn dray at Adnams' brewery tap, the Sole Bay Inn in Southwold, Suffolk

Lancastrian brewery Thwaites' shire horses on show

The Gipsies Tent, a former home-brew pub in Dudley, West Midlands

Shakespearian quotations outside Batham's brewery tap, the Vine ('Bull and Bladder') in the West Midlands

Early transport at Ruddle's in Rutland

Hook Norton — a classic tower brewery in Banbury, Oxfordshire

The cover of an early twentieth century Hall & Woodhouse price list

Donnington Brewery in the Cotswolds from across the millpond

The Simpkiss Brewery in Brierley Hill, West Midlands, in July 1985 – boarded up and derelict within weeks of takeover

Beard's brewhouse in Lewes, Sussex, disused since 1958 and in disrepair

The mash tuns at Davenport's in Birmingham

The mash tun at Harvey's in Lewes, Sussex

Fermenting vessels at Harvey's

Vigorously fermenting beer in Davenport's fermenting room

The miniature tower brewery at the Three Tuns in Bishops Castle, Shropshire

The changing face of brewing – a Matthew Brown dray asks for deliverance from the S & N takeover bid, but is itself delivering to a Theakston pub (Brown acquired Theakston in 1984). In the background is the brewery Mitchell's bought from Thwaites after the latter acquired Yates & Jackson in 1985!

McMullen's) and had been running it for some time. Nevertheless, despite the confidence expressed in Holden's future, Greenall's take-over of Simpkiss has cast a long shadow over the remaining independent brewers in the West Midlands.

Edwin Holden's great-grandfather, the first Edwin, began by brewing at the Brittania, a beerhouse behind the celebrated Old Swan in Netherton, in 1877, but he soon moved to the Park Inn in Woodsetton, then also a home-brew house. As the business expanded, the maltings behind the pub, owned by Atkinson's of Birmingham (later to be taken over by Mitchell's & Butler's), were bought and adapted to form the Hopden Brewery, undistinguished externally but clean and cleverly arranged inside, with a mixture of gleaming new fermentation vessels and second-hand plant from, amongst others, Mew Langton of the Isle of Wight, Cunningham's of Warrington and Hunt Edmunds of Banbury, none of them still in existence.

For a long time there were eleven pubs, many of them former home-brew houses in their own right, including the highly decorated, brassy Park Inn in front of the brewery. Since 1980, however, a deliberate policy of acquisition of other brewers' redundant houses has been followed. The latest additions are the Elephant & Castle at Pensnett, near Dudley, bought from Courage, and the Bull's Head in Sedgeley. All the tied pubs are to be found in the Black Country, including a particular favourite with many – offering all the traditional beers – the Cottage Spring, a bustling, comfortable pub in Wednesbury.

HOOK NORTON
The Hook Norton Brewery Co Ltd, Hook Norton, Banbury, Oxfordshire.
Pubs: 34. Beers: Mild (1032), Best Bitter (1036), Old Hookey (1049).

One of the most rewarding sights for lovers of real ale is that of a working brewery enhancing its environment, and there can be few more picturesque industrial views than that of the tower brewery at Hook Norton. Seen from across the fields to the south, the handsome late Victorian brewery blends perfectly into the otherwise pastoral scene. And after a period of uncertainty in 1982, after the death of Bill Clarke, the founder's grandson, the future of this exceptional country brewery seems reasonably secure – even though the vast majority of the voting shares are owned by the Gilchrist family, two of whom are also directors of the much larger, and ambitious, Burtonwood Brewery.

The first decisive step to establish the brewery took place in 1849, when John Harris, then aged only twenty-four, moved into a farmhouse in Scotland End in the Cotswold stone village of Hook Norton and set up as a maltster supplying small brewers in the immediate area. Within twelve months he had built his own brewhouse, with such success that he had to enlarge the maltings in 1865 and build a small five-quarter tower brewery, rising to three storeys, seven years later. By the time of his death in 1887 he had also acquired the first three tied houses, in Chipping Norton, Bloxham and Hook Norton itself. His successors were his son, John Henry Harris, and nephew, Alban Clarke.

The two businesses, of malting and brewing, proved so successful during the late nineteenth century, satisfying the thirsts of both agricultural workers (for whom a specially strong 'haymaker's brew' was produced at harvest time) and navvies engaged in railway construction, that a new brewery designed by William Bradford, a specialist brewery architect, was commissioned in 1900. It is a classic tower brewery, built of local ironstone, brick, slate, cast iron, weatherboarding and even some eccentric half-timbering, a fine and imposing piece of the living industrial history of the area.

In 1900 the firm of John Harris & Company became a limited company with the present title, though the ordinary shares were all retained by the family. 1917 saw the death in a cycling accident of Alban Clarke, the managing director, and consequently (since his son Bill was only thirteen) the appointment of the first director from outside the ruling family – Percy Flick, a Banbury estate agent. Although new to the trade, Flick was instrumental in repelling a take-over bid in the 1920s. Worse was to follow in 1950, on the retirement of Percy Flick, for although Bill Clarke had by then become head brewer, it became necessary to find new capital, and (to quote the official version), 'The brewery set out to strengthen its future by entering into a rewarding and amicable business association with the Gilchrist family who owned and operated a similar family concern brewing distinctive ales in Lancashire.' Today two of the three directors are Gilchrists (who own almost all of the voting shares and are also Burtonwood directors) but the management of the brewery is in the hands of David Clarke, great-grandson of the founder.

The beers produced in the brewery, still powered largely by its magnificent eighty-year-old steam-engine and boasting other equipment dating from 1900, with an unusual flat cooler and vertical refrigerator, no longer include the ale which was delivered to local farms by horse-drawn dray at a shilling a gallon in the early years of the twentieth century, but they are nonetheless worthwhile traditional brews. Credit for this perhaps belongs mostly to the brewing water, pumped from deep wells and so pure that it is used untreated. The local favourite is Best Bitter (1036), well flavoured, light and hoppy; mild (1032) is not very dark and not especially distinctive; Old Hookey (1049) is a fairly recent addition, a very pleasant dark brew with a traditional old-ale taste.

MARSTON'S
Marston, Thompson & Evershed plc, Shobnall Road, Burton-on-Trent, Staffordshire.
Pubs: 1,060. Beers: Capital Ale (1030), Mercian Mild (1032), Burton Bitter (1037), Merrie Monk (1043), Pedigree (1043), Owd Rodger (1080). There are also, at present, separate brews for the Border pubs acquired in 1984, though their future is doubtful.

There are a number of unusual features about Marston's. One of the biggest regional brewers these days, it clearly had pretensions to national status much earlier than the Big Six, for in the 1920s it collected brewers as far apart as Appleby-in-Westmorland and Winchester, but (possibly because it became becalmed under the Whitbread 'umbrella') it was strangely quiescent during the take-over free-for-all in the 1960s and has only recently re-entered the fray with the purchase of Border Breweries of Wrexham. Marston's pubs are therefore very widespread, and the free trade is even more far-flung. There are some remarkable beers, too: Merrie Monk is the strongest mild in the country, draught Pedigree is the only beer to be produced using the now-unique Burton Union system, and Owd Rodger is a really individual barley wine.

The firm began quietly enough, as J. Marston & Sons, with John Marston acquiring Coat's Brewery in 1825 and his bachelor son John Hackett Marston expanding the business until, in 1888, he retired and sold the firm to Henry Sugden. The merger with John Thompson & Sons took place in 1898, when the title Marston, Thompson & Sons was adopted. At the same time

Frederick Hurdle, whose great-grandson is the present managing director, became neutral chairman in an attempt to calm down feuds between the families. In 1902 the lease of the Albion Brewery, built in 1875 for the London brewers Mann, Crossman & Paulin, was acquired and three years later Sydney Evershed Ltd and its sixty-eight pubs was purchased, necessitating a further change in the company's name. So far the growth of the company had been achieved largely through the amalgamation of breweries within Burton itself, although breweries in northern Shropshire had been bought, together with the Midland Brewery in Loughborough, which brewed its own beers for thirty years after its take-over in 1901.

By the 1920s, however, Marston's ambitions were considerably greater, and a decade of astonishing growth ensued. In the north Bertwhistle's Appleby brewery was acquired in 1922; in the Midlands, the victims were Zachariah Smith's Trent Brewery in Shardlow, near Derby (eighty pubs, in 1922), George Pim of Stoke in 1925 and two more Shropshire breweries, in Market Drayton (the Crystal Fountain Brewery) and Newport. Further south a

Marston's logo commemorating 150 years of independent brewing

very significant purchase was that of the Winchester Brewery Company, bringing in 108 pubs in a completely new area in 1923. By now Marston's already had some 600 tied houses. The 1930s saw the Crown Brewery in Kettering and Lewis Clarke's of Worcester added, but the next major area of growth was the north-west, beginning in 1958 with the purchase of sixty pubs belonging to Taylor's Eagle Brewery in Greenheys, Manchester (whose brewery had actually been sold and dismantled in 1924). This was followed by the acquisition of Rothwell's of Newton Heath in 1961 (the brewery lasted seven years before it was shut down) and Smith's Crown Brewery in Macclesfield in the following year.

And that, for more than twenty years, was it. Marston's bought no more breweries during that period, perhaps because they themselves were now under Whitbread's crowded 'umbrella', having negotiated a trading agreement and share purchase in 1957. The Whitbread Investment Company now owns thirty-six per cent of the shares. Further growth did take place in 1984, with the purchase of Border Breweries of Wrexham and its 170 pubs. Ironically, Border had said only in the previous year that they were 'confident we can at least maintain our position and lay the foundations for steady progress. . . . Border Breweries will grow in strength in the next twenty years'. Within eighteen months the brewery – admittedly somewhat decrepit – had been closed (with the loss of five real ales) and, after an unseemly scramble in which Burtonwood Brewery's bid was shut out by Whitbread's manœuvrings (and other hopefuls such as Wolverhampton & Dudley were discouraged), Marston's emerged with a useful addition to their tied estate in North Wales.

Marston's therefore relies on its Burton brewery to supply a tied estate of over a thousand pubs, very widely dispersed and, with the free trade, serviced through depots at Penrith, Macclesfield, Rhuddlan, Wrexham, Worcester, Leighton Buzzard and Winchester. These depots offer a wide range of beers, six of them traditional draught but with three keg beers and two lagers as well. Starting at the bottom of the original-gravity range, Capital Ale (1030) is a very light-tasting brew, by no means well hopped, which performs well as a cheap light mild; Mercian Mild (1032) is a surprisingly full-flavoured and quite hoppy beer; and Burton Bitter (1037) is a fine staple bitter, very widely available and widely respected as a refreshingly bitter brew.

But the star of the Marston beers is undoubtedly Pedigree (1043), which began life as a bottled beer, Burton Best Pale Ale, and was converted to a draught premium bitter only in 1955, when production was a mere 400 barrels a week. Now more than 6,000 barrels a week are brewed. It is a pale-coloured brew with a very delicate flavour, deceptively so since it is a robust, full-drinking best bitter, widely and correctly considered one of the best in the country. It is also unique. This last attribute has only recently applied to Pedigree, for until the early 1980s Draught Bass was also produced using the Burton Union system, in which the beer ferments in oak casks and excess yeast is recirculated, further to stimulate fermentation. Although the method is costly, Marston's are committed to retaining their 900 Union casks – and their reward is a superb beer, subtly flavoured and perfectly balanced.

It is hard to believe that Marston's also brew two more outstanding beers, yet this is the case. Merrie Monk (1043) is a strong mild, the strongest in the country, and as its gravity suggests, it is not too dissimilar to Pedigree except in colour; rich and sweet, it is linked by the brewery to the medieval beers produced at St Modwen's Abbey. Owd Rodger (1080), first brewed at the Royal Standard of England in Forty Green, Buckinghamshire, is an exceptional barley wine, very rich, dark, sweetish and heavy – an excellent 'finishing' drink for an evening but dangerously addictive at any other time!

OLD SWAN (Ma Pardoe's)
The Old Swan Inn, Halesowen Road, Netherton, Dudley, West Midlands.
Home-brew pub. Beer: Home Brewed (1034).

One of the most difficult problems to resolve in writing about the Old Swan is what to call the beer. The Good Beer Guide describes it as 'bitter', which it patently isn't, but the most accurate description of it, 'light mild', is no more likely to be heard in the bar of the Old Swan itself than my compromise of 'home-brewed'. Regulars neither know nor care what to call it – they simply ask for 'a pint' and get on with the serious business of drinking it.

Those regulars received quite a jolt in 1984 with the death of Doris ('Ma') Pardoe and the subsequent decision of her family to sell the business, comprising the brewery, two pubs (the Old

Swan itself and the White Swan in Dudley) and off-licences in Halesowen and Sedgeley. Breweries were rumoured to be in the market for the pubs, and there were fears that the beer would disappear forever. Salvation was at hand, however, from an unlikely combination – Mercia Venture Capital, a West Midlands investment company, and CAMRA. The pub, brewery and recipe are now owned by Netherton Ales plc, with Mercia, CAMRA and 400 other shareholders having subscribed £275,000 to finance the new company.

The money will preserve (and sympathetically extend) a magnificent example of an unspoilt Black Country pub, bursting with character in its own right but also particularly important historically as the sole survivor of more than 200 home-brew pubs in this idiosyncratic part of the country. The public bar is extraordinary, with a white swan as the centrepiece of its green and white Victorian enamelled ceiling, engraved mirrors, an ancient stove and a weighing-machine; the smoke room is more restrained, with an austere elegance. And it sells good, cheap beer from the brewery behind the pub, as the Old Swan has done for more than a century.

The Pardoe reign spanned fifty-two years from 1932 to 1984, but the Old Swan itself dates from about 1840. In 1863 John Young

The Old Swan 'home-brew' logo

bought the pub, brewhouse, stables and gardens for £430, and in 1872 it passed to a Gloucestershire merchant, Thomas Hartshorne, in whose family the freehold remained until 1964. At the turn of the century it was known as the White Swan, but by 1910 it was the Old Swan again, declaring on the front page of *The Advertiser*: 'The Old Swan Still Swims. The Best and Purest Ales are now being sold at the Old Establishment known as the Olde Swan Inn, Halesowen Road, Netherton. Manager and Brewer, Mr Z. Marsh, late of Dudley. His Bitter is of an Extraordinary Quality brewed on the premises from Hops and Malt only of the finest quality, guaranteed absolutely pure.'

In 1932 Tommy Hartshorne offered the tenancy of the Old Swan to Frederick and Doris Pardoe, who had been married in 1918 and within a year were the licensees of the Bush at Dixon's Green, Dudley; in 1930 they moved to the British Oak in Sweet Turf, Netherton. At the Old Swan they appointed a new brewer, Solomon Cooksey (the Pardoes never brewed themselves, and indeed Doris was a teetotaller), who was succeeded by his son George largely because he refused to pass on the recipe and his brewing technique to anyone else. Even today the recipe is exactly the same as in 1932.

Frederick Pardoe died in 1952, and his widow Doris outlived him by thirty-two years, becoming a Black Country institution in her own right. Known as 'Ma' or 'Mrs' Pardoe to thousands of drinkers from all over the country, she served beer which, had she tasted it, she would surely have approved of. At its best (for it is not entirely consistent) it is a beautifully balanced pale, malty ale, with all the best characteristics of a light mild: a unique experience for beer drinkers. Its survival is little short of miraculous in the hard-nosed brewery world, and it is fervently to be hoped that Netherton Ales is true to its promise to retain the character of both the pub and its beer.

SIMPKISS
An Obituary Note

Simpkiss's four traditional beers disappeared almost overnight after take-over by Greenall Whitley in July 1985 had been followed by the immediate closure of the brewery. Never the prettiest of sights, on its former ironworks site in Brierley Hill, the Dennis Brewery was boarded up and derelict within weeks, and the fifteen pubs were selling Wem ales from Greenall's more

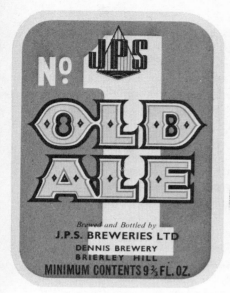 

Beers of the past: the Simpkiss brewery was closed in 1985

fortunate Shropshire subsidiary. With hindsight it is clear that the death in 1981 of Dennis Simpkiss paved the way for the sale of the family concern, with the loss of twenty jobs. The brewery workers learned of their fate through the Press; the company secretary was quoted as saying that 'There's been nothing from the Simpkiss family. It's an incredible state of affairs.' Yet in 1984 a bid from Greenall's had been rejected, chairman Jon Simpkiss saying that, 'Local people do like to drink local beers – and that's why we're staying in business.'

It was a sad end to a famous Black Country brewery, noted for its distinctive ales – particularly an excellent bitter and a unique, potent Old Ale which seemed to vary dramatically with each winter season. And it was the end of 115 years of typically West Midland small enterprise, from home-brewing at the Potters Arms in Delph through a period at the Home Brewery in Quarry Bank (Simpkiss's lost this brewery in a bizarre court case in 1916) to the building of the Dennis Brewery in 1934 and a partnership with Johnson & Phipps of Wolverhampton in 1955. Black Country beer can never be the same again: and all for the sake of fifteen pubs.

THREE TUNS
The Three Tuns Inn, Salop Street, Bishops Castle, Shropshire.
Home-brew pub. Beers: Mild (1035), XXX Bitter (1042), Castle Steamer (1045).

The Three Tuns is possibly the best-known and certainly the most imposing of England's clutch of historic home-brew pubs. Tucked away in a side street at the top of one of the country's smallest, sleepiest market towns, it has survived two changes of ownership since 1976 and seems set to go on brewing its popular, distinctive ales for some time to come. In doing so it will be slaking the thirsts not only of the local farm labourers but of a steadily increasing influx of beer 'connoisseurs', many regarded as regulars although their pilgrimages occur only once or twice a year.

The tower brewery across the yard from the pub may look laughably small compared with the brewhouse towers of its larger commercial rivals, but it has seen almost a hundred years of excellent home-brewing since it was built in 1888, and the pub itself has had a licence to brew beer since 1642. At the time the new tower was built, the pub had just come into the hands of the Roberts family, a situation which prevailed until 1976. In 1899 the quality of the ale was praised by no less an authority than the rector of one of the local villages, who said, 'I want them to exclaim with one voice, after they have tasted your beer, "Roberts deserves well of his country as he is the only man who has developed a cure for the agricultural depression!"'

At this time the beer was sold in quite a few local pubs and in cask to private customers, but this trade has largely been discontinued. Also ended is the practice of malting barley at the brewery itself; since 1935 best-quality Yorkshire malt has been purchased, although it is still milled on the premises and then hoisted to the top of the tower. The beer then emerges via mash tun, copper, hopback, cooler and fermenters, just as in much larger breweries, into the wooden or metal casks where it is dry hopped.

The Roberts' reign at the Three Tuns ended in 1976, when John Roberts decided to retire, finding that brewing combined with running a pub was too much of a strain. His successor was Peter Milner, who carried on for some six years, brewing once a week and selling twelve barrels or so in winter and perhaps twice as much in summer, before he too found the long hours too great a burden. In July 1982 Jim Wood, brother of the owner of Wood's

Brewery, the new brewery based in nearby Wistanstow, bought the business, and he now runs it with his sons and the Milners' brewer, Ernie Jones.

The three draught beers at the Three Tuns are headed by XXX (1042), a glorious strong bitter which is very pale indeed and extremely well hopped. The other long-established brew is mild (1035), dark, malty and refreshing. Peter Milner introduced the third beer, Castle Steamer (1045), a strong dark beer akin to porter, in 1979. The pub itself, a rambling multi-roomed listed building, is worth a visit in its own right, but the beer is still deservedly the main attraction.

WEM
The Shrewsbury & Wem Brewery Co. Ltd, The Brewery, Noble Street, Wem, Shropshire.
A subsidiary of *Greenall Whitley*, running 200 pubs. Beers: Pale Ale (1033), Mild (1035), Best Bitter (1038), Special Bitter (1042).

The Wem brewery, though it had formed part of the Greenall Whitley empire since 1951, survived the carnage of the late 1960s and early 1970s, which saw Greenall's concentrate all their other brewing activities at the Wilderspool Brewery in Warrington – closing breweries in Chester, Bolton, St Helens, Salford and Wellington in the process. The reason for Wem's survival would appear to be its profitability allied to the reputation of its beers – which are certainly the most distinctive now produced by a Greenall brewery. Recent years have seen expansion of the trading area, too, largely as an incidental result of Greenall's growth strategy: for example, Wem ales are now sold in the former Simpkiss houses in the Black Country, which were bought by Greenall Whitley in 1985.

Prior to its own take-over, the Wem brewery was perhaps most notable for having taken over a rival concern which was housed in a converted circus. The brewery was begun by William Hall and became the Shropshire Brewery Company in January 1898, at which time sixty-three pubs were owned. Later in 1898 the name was changed again, to the present title, when the business of Richards & Hearn in Bridge Street, Shrewsbury, consisting of ten pubs and a brewery, was acquired. This was the Circus Brewery, whose premises had originally formed a butter and cheese market and had later housed a circus. There were subsequent changes of use, too, for Wem closed the Circus Brewery in 1912, and it later became first a bakery and then offices before finally being demolished.

At the time of take-over by Greenall's in 1951, the Wem brewery was supplying ninety-four tied houses, but this number grew dramatically with the end of brewing at Greenall's other Shropshire subsidiary, the Wrekin Brewery in Wellington, in September 1969. By the late 1970s there were some 225 tied houses, but some of the less profitable pubs have since been disposed of and the number had fallen to 200, despite the opening of new pubs such as the Winning Post near Wolverhampton racecourse, before Greenall's push into the Black Country took real effect in 1985. The purchase of Ma Pardoe's second pub, the White Swan in Dudley, and a free house in Brierley Hill for which a staggering £1.7 million was said to have been paid, presaged the take-over and closure of Simpkiss, whose fifteen pubs now offer only Wem ales instead of locally brewed and very distinctive Black Country beer.

The great saving grace of the Wem Brewery was that it produced nothing but traditional ale – in stark contrast to the parent Warrington plant – but even this is no longer the case, kegging equipment having been installed in 1985. Ostensibly the keg beer is destined for the Midland club trade which Greenall's are keen to penetrate, but keg Special Bitter has been sighted in some tied houses, and some of the more rural pubs are threatened with conversion to keg beer only should their barrelage decline further. This would be an appallingly retrograde step, Wem having built their reputation on supplying traditional beer to the entire tied estate. The four beers are a very good, dry dark mild (1035), a very ordinary but still refreshing pale ale (1033), Best Bitter (1038) which is pleasantly hoppy and certainly the best of the Wem brews, and Special Bitter (1042), introduced in 1983 and quite well hopped.

8 The East Midlands and East Anglia

The Big Six run no breweries at all in this vast area, the last closure having been that of Watney's Norwich brewery in 1985. Their influence is undeniably strong, however, in areas such as Norfolk (where Watney's have a very strong local monopoly as a consequence of buying all the substantial breweries in Norwich) and Northamptonshire (again Watney's, who bought Phipps in 1960). Elsewhere there is a good deal of local competition, notably in Nottingham, where Home Brewery, Hardy's & Hanson's Kimberley Brewery and Greenall's subsidiary Shipstone still brew. Nottinghamshire also has the ambitious Mansfield Brewery, and Lincolnshire clings tenuously to Bateman's 'good honest ales'.

Several local and regional breweries supply East Anglia, notably Greene King, a major force with three breweries, at Bury St Edmunds, Biggleswade and Furneux Pelham, the last-named still trading as Rayment's. Adnams' Southwold ales are well known, and Suffolk also has Tolly Cobbold, a subsidiary of Ellerman Holdings but still producing local brews with an improving reputation. The other five East Midland brewers are a mixed bag. Everard's are perhaps the biggest, but Ruddle's are the best known, producing enormous quantities of processed beer for supermarkets together with draught County for Watney's pubs and the free trade. Elgood's are a much more traditional Fenland brewery, producing a bitter with a delightfully individual flavour. James Paine, taken over in the 1980s by a consortium of travel agents, now has only eleven tied pubs. Hoskins' of Leicester, also recently under new ownership, are the smallest of the lot, but not for long if present trends continue, since the original single tied pub has been joined by another four (one as far away as London) and a £2 million share issue is to be used to finance further expansion.

A number of excellent new wholesale brewers have started operations since the 1970s, including Woodforde's Spread Eagle Brewery in Erpingham, Norfolk, and Hoskins & Oldfield of Leicester, begun in 1984 by members of the family which used to run Hoskins' brewery. Their Hob bitter is an especially well-hopped and tasty beer.

ADNAMS
Adnams & Co plc, Sole Bay Brewery, Southwold, Suffolk.
Pubs: 68. Very considerable free trade. Beers: Mild (1034), Bitter (1036), Old (1042), Extra (1044).

Adnams' beers have had some fairly unkind things said about them in recent years, notably in 1977 (when yeast problems during brewery alterations caused some indifferent brews) and 1983 (again a result of brewery expansion). Even at their best, which is very good indeed, they are prone to being described as having a 'seaweed' taste, which seems a backhanded sort of compliment; the brewery did previously draw its water from a well which tapped a fresh water spring half a mile under the sea, but it was seaweed-proof and the beers are better characterized as very dry and hoppy.

The range of beers, served traditionally in all the tied houses

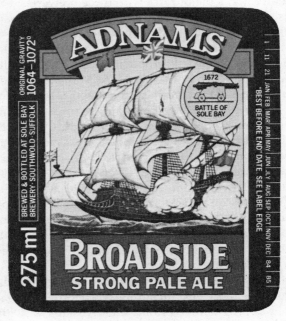

An Adnams beer label

and the vast majority of a thriving and widespread free trade, which accounts for over two-thirds of production, has recently been increased with the introduction of Extra (1044), a very fine best bitter which perfectly complements the 'ordinary' – as with Brakspear's. Extra is surprisingly dry-tasting and generously hopped for its gravity, which ensures a full, rounded flavour. The bitter (1036), despite accusations that it has become blander of late, is in fact still brewed to the same recipe and is superbly crisp, hoppy and aromatic. Dark Mild (1034), more robust than some of its counterparts, is sweeter and smoother and like many others is in some danger: it now warrants only one brew a fortnight, and sales continue to fall. Old (1042) is in a class of its own, a superb full-bodied dark winter ale. Tally Ho (1075) used to be available on draught around Christmas but these days is confined to bottles.

The sixty-eight tied houses include some real gems tucked away deep in the East Anglian countryside, although there are now some urban houses too, as Adnams have expanded into Norwich and Ipswich. Famed for their social conscience in respect of low-barrelage pubs, Adnams have in fact closed no fewer than fourteen pubs since the mid-1960s (eleven of them are now free houses, the other three no longer serve as pubs) whilst buying thirteen houses to maintain their tied estate at about the same level.

Most of Adnams' decisions lean heavily on the side of tradition, however. Traditional plant such as open fermenters has been retained in successive brewery extensions, and transport between the brewery and the nearby distribution centre (built to provide more space for brewing in the tiny, cramped site in the middle of Southwold) is by horse-drawn dray, which is also the method used to get beer to pubs within a five-mile radius. Adnams' animal world has expanded, too: the dray horses, Percherons, have been joined by 600 pigs which have possibly the most enjoyable feed in the country – a mixture of yeast, waste beer and meal. The yeast, incidentally, is a four-strain variety acquired from Morgan's a few days before their Norwich brewery was flattened by bombs in 1943 (Adnams' previous strain having, yet again, become infected!).

By 1943 Adnams, then as now brewing on the site of a brewhouse established by 1641, was already a well-established country brewery, having become G. & E. Adnams in 1870 and then been incorporated twenty years later as a means of raising the cash to rebuild the brewhouse. The company then expanded by taking over other small local breweries – Fisher's Brewery of Back Lane, Eye, in 1904, the tied pubs of Edward Rope (a brewery previously taken over by Flintham, Hall & Co of Aldeburgh) in 1922, and then Flintham, Hall themselves in 1924. It is still family-run, with John Adnams as chairman and Simon Loftus (whose grandfather purchased a substantial interest in 1902) the director in charge of Adnams' extensive wine-shipping interests. The company's outlook is determinedly independent and – providing the yeast can take the strain! – seems certain to continue brewing superb, seaweed-free beers for years to come.

BATEMAN'S
George Bateman & Son Ltd, Salem Bridge Brewery, Wainfleet All Saints, Skegness, Lincolnshire.
Pubs: 99. Beers: Mild (1032), XB (1036), XXXB (1048).

Bateman's, founded in the Lincolnshire Fenland in 1874, prospered quietly for over a century, producing magnificent beer sold mostly in small country pubs, until it was rocked to the foundations in 1985.

In the beginning, in 1874, George Bateman leased the Wainfleet Brewery from Edwin Crow, buying the freehold in the following year and then transferring to a brand-new site at Salem Bridge in 1876. This is the present brewery, next to a disused ivy-smothered windmill which was adapted to serve as the company's trademark for decades.

George Bateman's son, Henry, became chairman in 1921 and stayed in the job for forty-nine years, painstakingly building up the brewing side of the business and also buying a wine and spirits merchant, Ridlington's. Grandson George Bateman has been in the chair since 1970, during which time a new beer has been introduced and the company's very positive policy towards small rural pubs has been widely acclaimed. But in July 1985 George's brother and sister announced that their combined sixty per cent share of the business was for sale. As the predators closed in, George Bateman had the unenviable task of matching the highest offer.

The saddest aspect of all this was that any new owner would find it difficult to match George Bateman's enviable record of keeping open low-barrelage outlets if at all possible. Indeed,

Bateman's have a unique approach to the problem of supplying cask beer to their smaller pubs, supplementing the usual cask sizes with twelve- and six-gallon vessels and even wooden two-gallon casks known as 'piggins'. Many of Bateman's pubs are delightful country havens which are popular in the tourist season but in which trade falls by as much as two-thirds in the long winter months; only a quarter of the pubs sell more than four barrels a week (at which level of trade most big brewers would shut them down immediately). Bateman's philosophy is that there is a duty to serve small villages and hamlets which have no other facilities.

This is not to say that the firm lives entirely in the past, however. Their beers are available in the London free trade, and in 1984, after George Bateman had visited the Milan trade fair, an order was won to supply 23,000 bottles of strong pale ale to Italy every three weeks. Although this order increased production by eight per cent, the bulk business of the company remains domestic and traditional: the three real draught beers still account for seventy per cent of output. Bateman's 'good honest ales' are a dark and creamy mild (1032), outstanding at its best but, as it accounts for only a tiny fraction of production, variable in some outlets. XB, the ordinary bitter (1036) is also outstanding, very hoppy yet beautifully balanced. XXXB (1048), introduced in 1977 to cater for the premium bitter market, is more malty and has a fuller flavour.

As with all the best traditional breweries, all the beers are very drinkable, and the difficulty on entering a Bateman's pub is to choose one in preference to the others.

ELGOOD'S
Elgood & Sons Ltd, North Brink Brewery, Wisbech, Cambridgeshire.
Pubs: 55. Beer: Elgood's Bitter (1036).

The North Brink brewery, with its fine Georgian façade, has been producing Fenland ales since 1790. Elgood's sole remaining real beer, the draught bitter (1036) is popular and tasty, a really distinctive product which deserves to be more widely appreciated. Remarkably, it is available only in just over half the fifty-five tied houses in traditional form, top pressure being the favoured dispense system in the rest. This proportion of tied houses offering traditional beer is scarcely higher than when CAMRA was formed, and Elgood's must be one of the few breweries which (mainly, it seems, through inertia) have charted a virtually unchanged course over the last fifteen years. Top pressure is understandable (though not inevitable) in a few rural pubs which sell little beer, but half of Elgood's Wisbech pubs use the system, a surprising state of affairs.

Elgood's dark mild is yet another casualty of the trend to lighter beers, the draught version having been withdrawn in 1983. For years production of the mild had been sustained only by the fact that the same brew was used for the bottled brown ale, but demand for draught mild slumped so low that publicans found it impossible to keep in good condition. An excellent strong ale brewed to celebrate the Elgood centenary also failed to sell well enough to justify production on a permanent basis. The range of bottled beers has contracted, too, with the brewery's best and strongest, the rich and dark Old English Ale, also a victim of declining sales.

There are fifty-five tied houses today – three fewer than in the centenary year of 1978 and fifteen fewer than at the beginning of the Elgoods' reign – and they include notably unspoilt country inns, many of them catering for fishermen attracted by the Fenland dykes and drains. Especially good examples are the Black Hart at Rings End and the Four Horseshoes at South Eau Bank, Throckenholt – both with bitter served direct from the cask. The Boat at Whittlesey, a bustling town local, also deserves a special mention. There is very little free trade: Elgood's, possibly uniquely, don't supply free houses within the thirty-mile radius covered by their tied estate, and they don't supply beyond fifty-five miles in any case.

The brewery's policy is very much one of consolidation within the existing tied estate, and this seems to have been the case since Nigel Elgood's great-grandfather John Elgood mashed the first brew on 10 October 1878. Previous owners included John Cooch and Dennis Herbert, who converted an existing granary into the Georgian brewery complex which today stands almost unchanged; Thomas Fawcett, who bought the brewery and four pubs in 1795; and Colonel William Watson, historian and brewer, who with his partner Abraham Usill achieved the greatest expansion of the business, to forty pubs by 1836, when Phillips, Tibbits & Phillips took over. After this succession of changes the Elgoods brought a much-needed stability into the business, ensuring that

Fenmen of the 1980s can drink the idiosyncratic and thoroughly enjoyable draught bitter.

EVERARD'S
Everards Brewery Ltd, Castle Acres, Narborough, Leicester.
Pubs: 145. Beers: Old Original (1050), Old Bill (1068). Also market Everards Bitter (1035), brewed by Whitbread at Samlesbury, and Burton Mild (1033) and Tiger (1041), brewed at the Heritage Brewery in Burton-on-Trent.

The history of Everard's is remarkably complex, not because of a string of take-overs but because of a quite unparalleled series of changes of address. Thomas and William Everard, together with Thomas Hull, leased a brewery in Southgate Street, Leicester in 1849, promising their customers that, 'No effort shall be wanting in the production and supply of genuine ale of first-rate quality.' By 1896 they had also acquired the Bridge Brewery in the brewing capital of Burton-on-Trent. Only five years later they moved across town to the larger Trent Brewery (later renamed the Tiger Brewery), which had been built for the Liverpool brewer Thomas Sykes in 1881 and had originally been known as the Cripplegate Street Brewery. Two breweries were operated until 1931, when the Southgate Brewery was closed and converted into a distribution centre. Everard's were now in the unique position of having their only brewery situated outside their trading area.

The 1980s have brought still further changes, with the closure of the Burton brewery announced and all Everard's operations concentrated on a greenfield site on the southern outskirts of Leicester; the brand-new brewery there is a small-scale affair, however, designed only to brew Everard's low-volume, high-gravity ales. The intention was to contract out the brewing of the other beers, and indeed Whitbread now produce the ordinary bitter. However, an ambitious plan to preserve the Tiger Brewery in working order as part of a National Brewery Museum persuaded Everard's to buy in their mild and Tiger bitter from the new company, Heritage Brewery Ltd, formed to run their former brewery.

The dynamic nature of Everard's business in recent years does not end there. In 1978 they were the major beneficiaries when Ruddle's sold off their entire tied estate, buying twenty-four pubs in a move which was not entirely to the liking of the regulars, used to their excellent pint of Ruddle's. In an even more bizarre move, the Leicester brewers set up a brewery in the Falkland Islands in 1983 to produce Penguin Ale for troops and islanders alike (unfortunately the venture ended in failure in 1986). The recent past has seen substantial expansion into the free trade, a first tied pub in London (the Radnor Arms in West Kensington) and experiments with other brewers' beers in Everard's pubs, with a full-scale swap arranged in 1985 involving the shipping of Old Original to north-east England in return for Cameron's Best Bitter in Everard's own houses.

Everard's new brewhouse at Castle Acres (where the twelve-acre development as a whole cost £10 million) is an ultra-modern and highly automated plant which is computer-controlled and in its pre-fermentation stages consists of a multi-duty vessel and lauter tun. Its main task is to produce up to 15,000 barrels per annum of Old Original (1050), the rather sweet, malty and powerful bitter whose introduction in 1976 signalled Everard's return to the production of cask-conditioned beer after almost a decade in which the entire output was chilled and filtered. The plant also produces Old Bill (1068), a strong winter ale introduced in 1985 and named after Everard's extinct barley wine.

The contract brews include two which are unchanged from the days when Everard's themselves ran the Tiger Brewery. Burton Mild (1033) is a pleasantly dry and distinctive dark mild which perhaps ought to be better known, and Tiger (1041) a rather middle-of-the-road best bitter. But Everard's Bitter (1035), trunked down from Whitbread's Samlesbury brewery in Lancashire, is eminently forgettable and no substitute at all for the well-balanced and very drinkable Burton-brewed Beacon Bitter which it replaced. A CAMRA survey found that over ninety per cent of Everard's licensees would prefer to serve Beacon if that were possible, and almost as many felt that sales had reduced as a result of the switch. At least Everard's customers looking for a pint of ordinary bitter can opt for Cameron's bitter instead of the Whitbread brew.

GREENE KING

Greene, King & Sons plc, Westgate Brewery, Bury St Edmunds, Suffolk. Also at The Brewery, Biggleswade, Bedfordshire.
Pubs: 795. Beers: KK Light Mild (1031), XX Dark Mild (1031), Simpson's (1033), IPA (1035), Abbot Ale (1048), Christmas Ale (1052).

Also owns *Rayment*, which brews separately.

Greene King is one of the 'second division' brewers of Britain in terms of size: not one of the Big Six, it is equally emphatically not just a local brewery company. East Anglia's largest brewer, with three breweries and 800 pubs, also has an extensive free trade which accounts for about half the business. Yet, although it has adopted an aggressive posture at times (notably when Tolly Cobbold was rumoured to be up for sale), the company's reputation is more that of a sleeping giant, slow to adapt to change – hence the majority of Greene King pubs still use top pressure to dispense the otherwise excellent beers.

Greene King has an enthralling business history, written by Richard Wilson, which demonstrates both the early acquisitiveness of the company and the tenuous hold it has sometimes had more recently on its own independence. Benjamin Greene and W. Buck founded the original company in 1799; ninety years later came the merger with F. W. King & Son and the adoption of the present name. Take-overs came thick and fast after that. Amongst the first were the Newmarket Breweries in 1896, Clarke Brothers of Bury St Edmunds (the only local opposition) in 1917, and both Christmas & Company of Haverhill and Oliver Brothers of Sudbury in 1918. All these were in Suffolk, but Greene King soon outgrew its native county, mopping up the Bocking Brewery in Braintree in 1919 and moving into Cambridge with the acquisition of the Panton Brewery in 1925. Five years later the action moved north, with Ambrose Ogden & Sons' Sun Brewery in High Street, March, snapped up together with another forty pubs.

In 1938 plans for a merger between Greene King and Tollemache's Breweries of Ipswich were aborted only at the last minute (reputedly when Greene King looked hard at Tolly's accounts), but it was after the Second World War that fear of take-over really began to stalk the local independent brewers. At first this worked to Greene King's advantage: Simpson's of Baldock, whose fine Georgian brewery was demolished in 1968, asked to be taken over in 1954, and the Sudbury brewers Mauldon's approached Greene King for the same reason in 1958. Later that year Daniell's of Colchester did the same but were turned away – no doubt reluctantly – because Greene King hadn't the necessary capital, and within months Truman's had absorbed the 146 pubs.

The increasing scale of take-over activity involving the bigger brewers (including J. W. Green of Luton) began to concern Greene King in the late 1950s, and they even considered a defensive amalgamation with Simond's of Reading and Flower's of Stratford-upon-Avon. The possibility of one of the emerging national brewers taking an 'umbrella' stake was also investigated, and at one time Bass and Canadian Breweries held eleven per cent of the shares and Guinness a further eight (the latter still have five per cent). In the event Greene King's problems were solved by a merger in 1961 with Wells & Winch of Biggleswade, who had themselves taken over a number of local competitors and now controlled 287 pubs. The merger had two great advantages: it made Greene King too big for their local rivals to swallow, and because the Redman family owned most of the shares in the Biggleswade firm it consolidated the hold of the reorganised board (which these days still has two Redmans amongst its number) on the enlarged company. Ironically, Wells & Winch was one of the Whitbread 'umbrella' companies, but Greene King moved fast and the deal was sewn up before Whitbread learned of the idea.

IPA Bitter. Abbot Ale. XX Mild. BBA and KK Light Mild. What other brewery offers so many individual cask conditioned beers?

Renewed interest in traditional beers from Greene King: an advertisement from the late 1970s

Since 1961, however, Greene King has consolidated rather than expanded. There have been no more acquisitions and the three breweries (Bury, Biggleswade and the much smaller Rayment's brewery at Furneux Pelham; Simpson's brewery in Baldock survived take-over by eleven years but was closed in 1965) have been developed not to serve a larger tied estate but to cater for burgeoning demand from the free trade, which accounted for less than one in ten pints brewed by Greene King in 1970 but had grown to almost half ten years later. Most of the free trade is in the company's flagship beer, the strong Abbot Ale, but the biggest seller overall is IPA. Sadly (and predictably) sales of both the light and dark milds are declining fast and their survival is far from assured.

Abbot Ale (1048) is by far the best-known beer – indeed it is one of the best-known real ales in the country. Surprisingly it was not available as a draught beer until 1957, when a specially racked draught version of what was then a bottled pale ale won a major award. As a result Greene King tried draught Abbot in a few of their pubs, and after the beer won the Supreme Championship at Brewex in 1968 it became more widely available and sales rocketed. The beer is now sold in most tied houses, some Whitbread pubs and hundreds of free-trade accounts – and rightly so, for this robust, very full-flavoured darkish premium bitter is a superb drink, the more so since the brewery have resisted the temptation to reduce the hopping rate, with the result that for its gravity Abbot is a remarkably hoppy, aromatic bitter beer.

Clean-tasting bitter beers are the hallmark of Greene King, since IPA (1035) is also pleasantly well-hopped and dryish. It has the merit, too, of providing a sensible lunchtime alternative to Abbot, which because of its alcohol content suffers from what brewers quaintly term 'volume limitation'. Perhaps with this in mind Greene King introduced a new light bitter in 1985 – Simpson's (1033), which is available only as a cask-conditioned brew. The name is derived from the Baldock brewery closed in 1965, but the origin of the recipe is uncertain: it does seem to lack the characteristic bitterness of other Greene King beers.

Cynics have suggested that the new low-gravity bitter may be designed to replace one or both of the mild ales which are still brewed, albeit in small quantities. KK Light Mild (1031), brewed only at Biggleswade, is reasonably popular, but its dark counterpart, XX (1031), though brewed at both Biggleswade and Bury, is likely to be confined to the latter brewery before long. The much stronger Christmas Ale (1052) is the only dark beer with an apparently secure future; sales of this excellent, malty and not especially sweet brew have been impressive since it was reintroduced in 1980. Greene King also have a better reason than most for the name they have chosen for their winter ale, having, as we have seen, taken over the firm of Christmas & Co some seventy years ago!

HARDY'S & HANSON'S (Kimberley Ales)
Hardy's & Hanson's plc, Kimberley Brewery, Nottingham.
Pubs: 206. Beers: Best Mild (1035), Best Bitter (1039).

The story of the Hardy's & Hanson's breweries, which eventually merged in 1930, starts with two concerns brewing separately but on adjacent sites in the village of Kimberley. Stephen Hanson was the first to brew, building his own brewery in the village in 1847; William and Thomas Hardy, brothers who were wholesale beer merchants, began by taking over the established brewery (actually a converted bakehouse) of Samuel Robinson a decade later. Within four years the Hardys were ready to expand, building a new and much larger brewery 200 yards from their original site – and immediately across the street from Hanson. At this time both brewers offered mild ale in four strengths, from X Mild to XXXX Strong, together with a light dinner ale and special pale ale.

Both the Hardys and the Hansons had been drawn to Kimberley by the excellence of the brewing water, but maintaining the supply from the Alley Spring and the Holly Well taxed their ingenuity to the limit. In the 1870s the Hardy brothers tried to gain monopoly rights to the spring water, which would have put Hanson out of business, but in 1874 circumstances forced the two breweries to co-operate in carting water from a nearby colliery when their supply failed. Two years later they united again to beat off a threat from the Midland Railway, who wanted to dig a cutting blocking the watercourse but were eventually persuaded to build their new line in such a way that water supplies were safeguarded.

In 1861 the sudden death of Stephen Hanson almost spelled the end for the fledgling brewery, but his eighteen-year-old son kept the business alive. Sixty years later the Hardys were in even graver difficulties, with no heir and with beer described as

A Hardy's showcard from the 1920s

cloudy, lacking character and possessing a purgative effect. Eben Hardy's sudden death in 1926, together with falling sales, precipitated talks of a merger with their rivals and neighbours, who were well supplied with sons and heirs ready to take on the brewery. The discussions which led to merger in October 1930 were a closely guarded secret, for neither side wished to see rival bids from larger breweries. The two firms continued to trade separately until 1972, although Hanson's brewhouse, superbly designed but smaller than Hardy's, had closed forty years earlier. Customers refused to accept for many years that there was no difference between Hardy and Hanson beer, even to the extent that casks labelled with the favoured name had to be delivered to certain outlets.

Until 1965 all the beer was traditional draught beer, racked into casks, but chilling and filtering equipment was then installed, mainly to serve the club trade. In 1970 the company went so far as to introduce Kimberley Keg, but this sweet keg bitter lasted only eight years before being withdrawn. Some Kimberley pubs do, however, dispense beer by top pressure or under a light blanket pressure. Other changes occurred in the 1970s: the Hardy maltings, built in 1875, were closed in 1973 because of difficulty in recruiting skilled maltsters, and Hanson's brewhouse, disused for forty years but kept standing for sentimental reasons, was demolished in 1973. The first executive director from outside the two families joined the board in 1974 – although the families still have a considerable, though not controlling, interest in the company's shares. Finally, a major brewery expansion scheme funded from the company's own resources began in 1979.

The 1970s had also seen a gradual reduction in the bitterness of Kimberley's draught beers, and certainly the Best Bitter (1039) is a rather sweeter brew than some of its East Midland rivals. Best Mild (1035), however, is a full, dark and malty beer highly regarded by many mild drinkers. PMA, a light mild restricted to a small number of outlets, was discontinued in 1982 when production had fallen to only a few kils a week. The bottling line, too, has gone, though the beers are still brewed at Kimberley and then transported in bulk to be bottled under contract. Guinea Gold, a low-gravity light ale, recalls Hanson's Guinea Ale, a popular family ale described as 'possessing rare tonic properties' in the 1880s and named from Robert Hanson's trademark – a George III golden guinea.

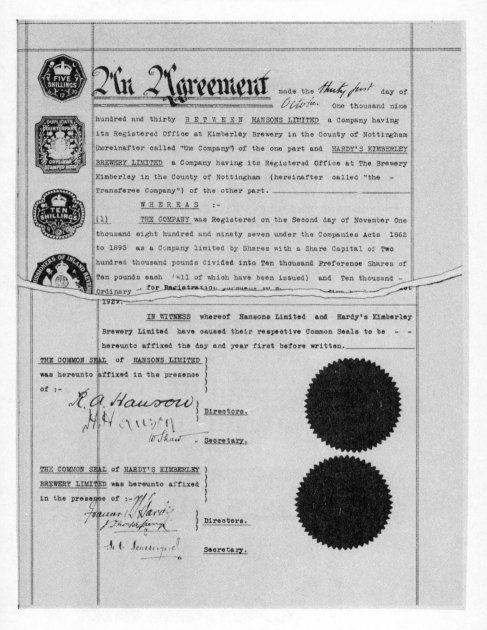

The 1930 merger agreement between Hardy's and Hanson's

HOME ALES
Home Brewery plc, Mansfield Road, Daybrook, Nottingham.
Pubs: 449. Beers: Mild (1036), Bitter (1038).

Home Brewery, despite the name and despite the low profile it has consistently adopted, is one of the largest of the surviving local brewery companies. There are 449 on-licences, including eleven hotels, and a thriving free trade in pubs and clubs throughout the East Midlands and, more sparsely, from Cambridgeshire to South Yorkshire and from Birmingham to the Lincolnshire coast. And Home Brewery's prices are notably low, primarily as a result of fierce competition between Nottingham's independent breweries. Yet Home have carried out an extensive modernization programme, with the Daybrook brewery substantially rebuilt since the early 1970s and a new warehouse and beer processing block added in the 1980s.

The Daybrook site was originally used as a maltings by two brothers, John and Samuel Robinson, who supplied local publi-

A Home Brewery beer label

can-brewers. Demand from local pubs encouraged them to switch to brewing in 1878, and by 1890 the venture had proved so successful that a limited company was formed, adopting the title 'Home Brewery Company' from the Robinson family's Home Farm estate in Daybrook. Several home-brew pubs were bought, but major expansion required brewery acquisitions, the most important of which were Hutchinson's Prince of Wales Brewery in New Basford, Nottingham, in 1914 and George Green in 1922. These purchases brought more than a hundred pubs into the company's hands, but Home Brewery's impressively large tied estate nowadays is due more to patient pub-by-pub expansion, as with the 1984 agreement which saw Home take over twelve pubs from Allied Breweries in return for taking Skol lager. Even more remarkably, in the twenty-five years after building-restrictions were lifted in 1952, Home Brewery opened more than a hundred new pubs.

This conscious policy of piecemeal expansion has required considerable investment in the brewery, beginning with a totally new brewhouse, complete with push-button control panel, in 1975. The brewhouse is capable of being extended to produce up to 20,000 barrels of beer a week, more than twice the current output, begging the question of whose pubs the surplus capacity might be ready to supply. A temporary fermenting block, with a mixture of traditional open and modern conical vessels, was the next stage of the redevelopment, followed by a £23 million scheme in 1983–6 to double capacity by providing a beer warehouse and a hundred-foot-high beer-processing block.

Despite the new equipment and technology, two-thirds of Home Brewery's output is still cask-conditioned beer. All the beers are derived from two basic brews, mild and bitter, and the keg and bottled beers are derived by parti-gyling these two brews – that is to say, dividing and diluting the basic wort to make the required quantities and qualities of finished beer. The two traditional brews, available from handpump or electric pump in the majority of the tied pubs, are mild (1036), which is a dark, creamy, quite high-gravity and good value brew, and bitter (1038), lightly hopped and well balanced.

The beer warehouse, opened in 1983, includes a temperature-controlled beer store, covering 24,000 square feet, where casks of traditional beer are stored on pallets for seven days before being weighed (Home's cask beers, rather unusually, are sold by weight rather than volume) and loaded onto drays. It is all very different from the early days, when the casks were stored in cellars tunnelled into the sandstone rock some ten feet beneath the brewery. The *Nottingham Daily Express* reported in 1882 that John Robinson's cellars 'are not to be equalled in the Midland counties', with 'storage for more barrels of fine, pale, bitter and strong ales than one would care to commence counting'.

HOSKINS'
T. *Hoskins Ltd*, Beaumanor Brewery, Leicester.
Pubs: 5. Beers: Best Bitter (1039), Penn's Ale (1045), Old Nigel (1060), plus occasional special brews.

Hoskins' is very much a brewery which has seen a dramatic change in its fortunes in the 1980s; for years the brewery served just one pub, fifteen miles away from the brewery in Market Bosworth, and a couple of off-licences, and the beers, though highly regarded by some, were a little too variable for comfort. Now Leicester's only surviving long-established brewery has new owners, and an era of expansion is well underway.

The firm's origins can be traced back only as far as 1890, when Jabez Penn, a Warwickshire man who had come to Leicester as a blacksmith and grocer a decade earlier, began to brew beer on Beaumanor Road. The original brewhouse was built in 1895 and extended three times in the next fifteen years, doubling its original capacity. By 1901 Thomas Hoskins, Penn's son-in-law, had been taken into partnership, and within five years he had assumed full control of the business; the 1906 balance sheet, valuing the entire business at £898, still exists.

Tom Hoskins ran the brewery until the 1950s, winning a series of medals at brewers' exhibitions in the 1920s and 1930s; his son George continued the family tradition until his death in 1976. The family persevered for another seven years but finally put the brewery, by then loss-making, up for sale. The new owners, Barrie and Robert Hoar, who then ran the Saffron Walden Vineyard and Cyder Orchard Company, took control in July 1983, when the legal documents were signed on an upturned case balanced on a high stool (there being no table in the brewery).

Major changes have quickly taken place in the brewery, the beers and the tied estate. The brewery has been spring-cleaned, bottling has been discontinued because of the state of the equip-

A Hoskins of Leicester advertisement

ment (IPA is still bottled, but by Holden's), and a new boiler has been installed, saving up to sixty per cent in energy costs. The range of beers has been radically overhauled. The Best Bitter (1039) survives, although it is more consistent and seems a little darker and sweeter than before, possibly through the addition of crystal malt. Gone is the mild, a victim of falling demand. New beers are Penn's Ale (1045), a dark and full-bodied strong bitter named after the founder, and Old Nigel (1060), a very potent sweet brew. Other brews celebrate special occasions.

Even more far-reaching have been the changes to the tied estate. The original tied house, the Red Lion at Market Bosworth, has been retained and partly modernized, and four others have so far been added. 1984 saw the opening of the Tom Hoskins at the brewery itself and of the Crown & Cushion in Whitwick, acquired from Ind Coope. The Rainbow & Dove in central Leicester (also formerly with Ind Coope) and the Waterside Inn in London N1, bought for a staggering £445,000, were opened in 1985.

The pubs are run to a certain extent as free houses, taking other brewers' real ales to sell alongside the Beaumanor Road products. The Tom Hoskins, indeed, had offered beers from fifty other brewers within a year of its opening. But the main attraction in the pubs is undoubtedly the chance to drink Hoskins ales, rescued from possible extinction in 1983 and now seemingly secure – the more so since the success of a £2 million share issue in 1986 (leaving the Hoar brothers with a stake of some thirty per cent), aimed at allowing further acquisition of tied houses.

MANSFIELD
Mansfield Brewery plc, Littleworth, Mansfield, Notts.
Pubs: 420, together with an extensive free trade in pubs and, especially, clubs. Beer: 4XXXX bitter (1045).

Mansfield Brewery is a comparatively little-known company, yet one which is fast-growing and ambitious, having taken over North Country Breweries in 1985 to double its tied estate – just seven years after a similar bid for Shipstone's was rejected in favour of Greenall Whitley's offer, and five years after Mansfield was transformed into a national soft-drinks distributor with the take-over of the Evesham firm of T. W. Beach, a business which accounts for over a third of the group's turnover, supplying the major national grocery chains with their own-label requirements. Mansfield's profits in 1985 were £8 million (comparable to those of Matthew Brown) and were expected to reach £11 million in 1987.

The brewery was founded in 1855 and has until recently enjoyed a comparatively placid existence under the control of the Chadburn family – William Jackson Chadburn having been taken into partnership in the 1860s and his descendants Robin and Richard currently serving as chairman and company secretary respectively. The largest outside shareholder, appropriately enough in a major coalfield area, is the NCB Pension Fund, with six per cent of the shares. Until 1978 and the abortive Shipstone bid, the only acquisition was that of the Chesterfield Brewery, way back in 1934.

The acquisition of T. W. Beach in 1980 was followed by an ambitious redevelopment of the brewhouse, completed in 1984 at a cost of £4 million. Built next to the existing fermenting rooms, the new brewhouse was designed to increase capacity by fifty per cent, with four 500-barrel liquor tanks, computer-controlled mash tuns, three coppers and two whirlpools capable of brewing 11,000 barrels a week. Within a year there was a pressing need for the extra capacity, for in an extraordinary £42 million deal Mansfield bought the Hull-based North Country Breweries from its parent company Northern Foods, which had decided to close down the Hull brewery, said to be old and inefficient.

The benefits for Mansfield Brewery were a doubling of the tied estate, to about 420, and an increase of about 400 in free trade outlets. And the trading area of the new acquisition, concentrated in Humberside and south Yorkshire, was largely complementary to Mansfield's own area of operations. The benefits to the beer-drinker were less easy to discern, since North Country's only real ale, Riding Bitter, was scrapped along with the rest of the Hull brews. Nevertheless, beer sales in Hull pubs had risen by twenty per cent within a year of the take-over.

The one traditional draught beer brewed at Mansfield is 4XXXX bitter (1045), first produced in 1982 and brewed in Yorkshire squares from a grist containing eighty per cent malt, with Fuggles hops. Although it is dry-hopped it is a full-flavoured malty brew, worth a try if it can be tracked down. Initially the beer was test-marketed in fourteen pubs, but although it is now available to any of the 420 pubs, providing they have suitable cellar accommodation, it is still not widely available (most Mansfield pubs, in fact, have temperature-controlled cellars which are too cold for cask-conditioned beers). At least a real draught beer is now produced – for nine years Mansfield produced none at all. Cask-conditioned beers were phased out in 1973, because the clubs were said to demand bright beer, and even in 1982 Mansfield stated that, 'We still believe the mass market needs bright beer.'

Mansfield's bright beers are a bitter (1039) and a mild (1036) which still accounts for a sizeable minority of production. There are still those, however, who recall the tastier, hoppier brews of the pre-1973 era. Marksman lager, relaunched in 1981, and Export lager are the other Mansfield brews, and they have already achieved successes in competitions. Mansfield also make the beers available in PET bottles (one-, two- and three-litre sizes) which are manufactured at the brewery.

PAINE'S

James Paine Brewery Ltd, The Counting House, Market Square, St Neots, Cambridgeshire.

Pubs: 11. Beers: XXX Bitter (1036), St Neots Bitter (1041), EG (1047).

In 1831 James Paine acquired the site of the brewery which now bears his name, having originally brewed beers for sale in his own home, Toseland Manor near Paxton in Huntingdonshire. In the next 150 years the business continued as a small-scale country brewery, with few tied houses and, in the early years, an unusual sideline which saw the brewery staff assisting local farmers and country gentlemen with harvest and Christmas ales brewed in their own kitchens, cellars and even private brewhouses. More ominously, the business was increasingly overshadowed by Paine's other businesses of home-brew kit manufacturing, malting and flour milling; by the 1970s the brewery represented less than a tenth of Paine's turnover, had no influence at board level and was starved of increasingly necessary investment. Worse still, the beer was variable and had an unenviable reputation, and production sank to a mere forty to fifty barrels a week at times.

It was scarcely surprising, then, that the take-over of the brewery business for about £2 million by a consortium of travel agents in 1982 was greeted with enthusiasm. The renamed James Paine Brewery quickly received much-needed attention, with a new boiler and new equipment to treat the town water used for brewing, and a programme of pub refurbishment was begun. Sales increased by more than half within a year, and the beer was widely regarded as much improved. Trading agreements with Charrington, Ind Coope, Tolly Cobbold and Matthew Brown increased production and widened the area in which Paine's products were sold. As if to underscore the emphasis on expansion, Paine also bought the London beer wholesaling business of Robert Porter from Gibbs Mew (whose involvement with Porter's had been generally unhappy) for about £500,000 in 1985.

But these gains had their costs, not least in a dramatic reduction in the tied estate. James Paine quickly sold four of the twenty pubs they had inherited in 1982, three as free houses still committed to selling the brewery's beers and one to a property developer. Despite the brewery's insistence that no more sales were envisaged, and that on the contrary purchases were anticipated, another five pubs had gone within three years.

There are three beers which, though they are criticized, can be sampled in traditional form. The recipes are still reputedly based on the nineteenth-century originals, although the beers themselves are considerably weaker than was then the case. The staple bitter, accounting for two-thirds of bulk beer production, is XXX Bitter (1036), mellow and without any distinctive characteristics. The others are St Neots Bitter (1041), previously known as '41' after the original gravity, and EG (1047), an unusual, sweet and malty beer. The initials are those of the Eynesbury Giant, James Toller, who died in 1818 and is buried in Eynesbury parish church, conveniently close to one of Paine's remaining pubs, the Hare & Hounds.

RAYMENT'S
Rayment & Co Ltd, The Brewery, Furneux Pelham, Buntingford, Hertfordshire.
A subsidiary of *Greene King*, with twenty-five pubs and a very large free trade. Beer: BBA (1036).

Rayment's brewery had a very short lifespan as an independent entity – a mere twenty-eight years from 1860 until its purchase by two Greene King directors – but almost a century later it is still in production and indeed appears to be increasingly prosperous. All bottled beers, all keg beers and indeed all but one draught beer have been discontinued in relatively recent times, leaving Rayment's to brew what they brew best – BBA, a popular and refreshing brew available in traditional form not only in the vast majority of the tied estate but in a very large number of free houses in north London and in parts of Hertfordshire and Essex.

William Rayment began brewing on a small scale (his brewing-liquor coming from a pond in the grounds) at Furneux Pelham Hall in the 1840s. His lease on the hall expired in 1860, by which time he was in business as a brewer, maltster, miller, farmer, brickmaker and drainpipe-maker, and in order to continue in brewing he built the present brewery at Barleycroft End, using bricks baked in one of his own kilns. Rayment died in the 1860s but the business was carried on by his executors with Stukeley Abbott as brewer and manager until 1888, when it was put up for auction. The buyers were two Greene King directors, Edward Lake and Frederick King, but they bought the concern privately, forming a private partnership to run the brewery as a 'family' business.

Labels for Rayment's bottled beers – a vanished breed

The partnership was dissolved in 1912, and Alan Lake, who had recently succeeded his father as managing director, controlled the brewery until it was finally taken over by Greene King in 1928. Before surrendering its independence, Rayment's experienced take-over from the other side, buying the Little Hadham Brewery Company in 1912 (and gaining just one tied house in the process). Since 1928 the Furneux Pelham arm of Greene King has been allowed surprising independence, partly because of its relative insignificance but also because of its impressive profitability: it had become the most profitable part of the GK empire by the end of the 1950s.

The brewery now, although its outward appearance has changed little, is an amalgam of very old, second-hand and very new. The old includes the brass-bound oak and copper mash tun, installed by Briggs of Burton-on-Trent in 1870, and a masher dating from 1920; both, unusually, are sited on the ground floor of the brewhouse. Amongst the second-hand items are a cask-washer bought from Banks's and a pre-1914 cask beer racking tank, formerly with Simpson's of Baldock but moved to Furneux Pelham when the Baldock brewery was closed – by Greene King – in 1965. The new comprises extra fermenting vessels to cope with demand for BBA and a new warehouse built in 1981.

Rayment's first bottling machine, installed during the First World War, was hand-operated and could deal with only six bottles at a time. Upgraded during the 1920s, the bottling line was shut down in 1973, with the loss of Rayment's bottled beers, notably Pelham Ale. The repercussions of this decision were far-reaching, however, for the end of Pelham Ale eventually condemned draught AK, which had been brewed by Rayment's since the 1890s and was highly regarded as a delicately flavoured mild ale.

The one beer still produced at Furneux Pelham, Rayment's BBA (1036), is, however, a particularly good bitter, and one for which demand has grown steadily. In 1970 Rayment's were brewing only two or three times a week, but by the end of the decade five brews a week was the norm, and indeed one reason for the demise of AK was the need to find extra capacity for the more popular bitter. BBA (Best Burton Ale) is a quite delicate but well-balanced brew produced from a grist containing almost ninety per cent barley malt, with maize and a little sugar added. Almost all the tied houses serve the beer by handpump (in the late 1970s less than half did so), and the free trade, which is of overwhelming importance to Rayment's, taking well over eighty per cent of production, also much prefers the traditional product.

RUDDLE'S
G. *Ruddle & Co plc*, The Brewery, Langham, Oakham, Rutland.
Pubs: none. Beers: Rutland Bitter (1037), County (1050).

There are two Ruddle's stories to tell: that of the 'old' company, founded in 1858 and peacefully existing as a country brewer supplying excellent ales to a small tied estate, and that of the 'new' Ruddle's, still committed to quality beers but selling most of their output in supermarkets and the rest in a widespread free trade, having sold the tied houses to finance expansion.* The way of life chosen by the new company is high-risk and, in the supermarket trade, low-margin, but Ruddle's profits broke through the £1 million barrier in 1984 (having been just £191,000 five years earlier) and production is still rising.

The Langham Brewery was built by Richard Baker in 1858 and inherited by his second son, Edward, three years later. The Bakers were also substantial landowners and employed an agent, James Harris, to manage the brewery. In its early days the brewery passed through several hands and by 1876 was owned by George Harrison, who also had brewing interests in Leicester. Within five years Boys & Style were running the brewery, and by 1895 H. H. Parry was the owner. George Ruddle became Parry's brewery manager in 1896, and when the brewery was sold after Parry's death in 1910, largely because he had no sons to inherit the business, Ruddle was the successful bidder, paying £19,500. The Ruddle family, originally farmers on Salisbury Plain, now had two breweries, since George's brother ran a small brewery in Bradford-on-Avon; this was, however, taken over by Usher's of Trowbridge in 1924.

For his money George Ruddle picked up the brewery, thirteen freehold pubs (many owned until 1978, an interesting exception being the Duke of Northumberland in Leicester, a city in which Ruddles have had no tied houses for decades), three leasehold pubs and six off-licences (again mostly in Leicester). Little change occurred at the brewery, and the tied estate grew slowly until 1957, when a new bottling hall was erected with money borrowed from Whitbread – marking the start of Ruddle's two decades under the 'umbrella'.

1978 saw major upheavals which heralded the arrival of the 'new' Ruddle's (under a new managing director, Tony Ruddle), with diversification into packaged beer, the rapid growth of free trade, the buying-back of Whitbread's thirty-one per cent shareholding, and the sale of the thirty-eight tied houses, which by then represented only fifteen per cent of the business. Everard's bought twenty-four of the pubs, Melbourn's of Stamford (no longer brewing but still owning pubs) and North Country Breweries five each, Samuel Smith's two, and Whitbread – despite Tony Ruddle's undertaking that no pubs would be sold to national brewers – the final two.

Cash raised from the sale was used to finance massive expansion of the brewery, with fermenting capacity raised to 1,750 barrels a week and more bottling and packaging machinery introduced to cater for the supermarkets. Packaged beer, much of it for Sainsbury and Waitrose, accounts for two-thirds of production; in the pub world Ruddle's is now available in hundreds of Watney houses in London and East Anglia, some Hall's (Allied Breweries) pubs and a vast number of free houses. A £3 million expansion, begun in 1983 and financed entirely from Ruddle's own resources, had the target of increasing capacity to no fewer than 5,000 barrels a week. The risk that Watney or the supermarkets will one day look elsewhere for their beer is one which is acknowledged: 'But the reason the supermarkets stick with us, when they could find beer cheaper elsewhere, is that they couldn't get beer as good elsewhere.'

Quality is of critical importance to Ruddles, then, and this certainly shows with the draught beers. Ruddle's County is the only beer to have twice won the Supreme Championship at the brewers' exhibition, Brewex, – in 1952, as a small country brewery, and 1980, as the present free-trade concern. County (1050) actually used to be a mild ale in the 1930s but after a break in production reappeared as a premium bitter in 1950. A rich, dark, full-flavoured beer which is heavily hopped to counteract the malty sweetness common in such high-gravity brews, it is one of the finest and most distinctive beers in the country. The other draught beer from Ruddle's, Rutland Bitter (1037), was increased in gravity by five degrees in 1985, a move which regrettably has lost the light, refreshingly hoppy taste of the beer, replacing it with a beer which is in some respects just one more 1037 bitter.

*In July 1986 Ruddle's shocked the industry by announcing an agreed take-over bid from Grand Metropolitan, though the future of the brewery was guaranteed.

SHIPSTONE'S
James Shipstone & Sons Ltd, Star Brewery, New Basford, Nottingham.
A subsidiary of *Greenall Whitley*, with 270 pubs. Beers: Mild (1034), Bitter (1037).

When Shipstone's finally sold out to Greenall Whitley in 1978, there were many who were pessimistic about the future of the brewery. After all, Greenall's had already shut well-regarded breweries such as Groves & Whitnall and the Chester Northgate Brewery and already supplied over a thousand pubs from their Warrington base. In the event Shipstone's are still brewing, their local identity remains and cash has been injected into the tied estate. But the cost has been enormous, for in the opinion of many drinkers Greenall's have ruined one of the great cask bitters of Britain.

An advertisement for Shipstone's bitter

In 1977 I wrote of Shipstone's 'celebrated and distinctive beers' and in particular the 'superb light-coloured bitter'. This was a marvellously clean-tasting, very hoppy, aromatic bitter which, pretty soon after the take-over, became a great deal less distinctive. Had the recipe changed? The brewery were adamant late in 1979 that the beer was 'the same as it has always been, and will remain that way'. Disgruntled drinkers continued to complain that a once-prized product was now a bland and mediocre offering, and in May 1981 Shipstone's admitted that the recipe had been changed by reducing the hop rate. This was reputedly the result of extensive market research but it has produced a bitter (1037) which is bland, uninteresting and not just pale but a pale shadow of its former self. In contrast the mild (1034), a quite dry, dark drink, is still worth sampling.

Greenall's have not stopped at the bitter recipe, either. Shipstone's prices have risen above those of their local competitors, and unprofitable pubs, together with city-centre pubs such as the Crystal Palace in Nottingham, worth more as a potential shoeshop than a pub, have been unloaded. Well-liked traditional pubs, such as the Windmill, also in Nottingham, have been converted into fun pubs. The Windmill became the Shoulder of Mutton, with live music, disco dancing, strobe lighting and an entry charge. In other pubs the staff are expected to perform a song-and-dance act on the bar between pulling pints. All this must have James Shipstone, who founded the brewer in 1852, whirling in his grave.

More traditional methods of expansion saw the Carrington Brewery acquired in 1898, the Star Brewery itself substantially rebuilt in 1900 and the Beeston Brewery swallowed up in 1922, with the former brewery converted into maltings which are still in use. The purchase of George Hooley's Wheatsheaf Brewery in 1926 added another thirteen pubs, and the final purchase, the neighbouring New Basford brewery of Thomas Losco Bradley Ltd, took place in 1954.

By 1978 Shipstone's were, by their own admission, short of cash, though there was no financial crisis. Northern Foods made a £13 million bid which the board described as 'wholly inadequate', and it was accepted by only a derisory four per cent of the Nottingham brewer's shareholders. But it soon became clear that the company was up for sale at the right price; Mansfield Brewery's £17 million was turned down, but Greenall's £20 million secured the brewery (with its spare capacity – only about two-thirds of the potential 5,000 barrels a week was being brewed) and around 280 pubs.

TOLLY COBBOLD
Tollemache & Cobbold Breweries Ltd, Cliff Brewery, Ipswich, Suffolk.
A subsidiary of *Ellerman Holdings*, who also own *Cameron's*.
Pubs: 340. Beers: Mild (1031), Bitter (1034), Original (1037), Old Strong (1046).

Tolly Cobbold's history is long and fascinating. Following the take-over in 1977 by Ellerman (themselves bought by the Barclay brothers six years later), what had appeared to many to be a slow decline seems to have been arrested, with more and better beers, brewery redevelopment and pub improvement, and an expanding trading area. Despite the two shifts in overall control, both the Cobbold and the Tollemache families are still represented on the board, ensuring for the present a local flavour in some of the company's activities.

The first brewery associated with the two families was built in 1723 by the established maltster, Thomas Cobbold, in King's Quay Street, Harwich. It was his son, also Thomas, who recognized the benefits of supplying the growing industrial town of Ipswich, and brewing was transferred to the Cliff Brewery there in 1746. Also transferred from Harwich were some of the original vessels, and one of the coppers is still in use today. The Cliff Brewery was completely rebuilt in 1896, and in 1923 the pubs of the third Ipswich firm, Catchpole's Unicorn Brewery, were bought jointly with Tollemache's Breweries when Catchpole's was voluntarily wound up.

The Tollemache family, though established in Suffolk since the eleventh century, did not enter the brewing industry until 1888, when they bought Cullingham's brewery in Upper Brook Street, Ipswich. They proved adept at their new business, however, moving further afield with the purchase of the Essex Brewery in Walthamstow in 1920, a wine and spirits merchants in Norwich in the following year, and a majority interest in the Star Brewery, Cambridge, and its 119 pubs in 1934. Just before that, the first thoughts of a merger with Cobbold's appear to have surfaced, but the amalgamation did not take place until 1957. The Cliff Brewery was enlarged following the merger, and the Upper Brook Street brewery finally closed in 1961, but it was not until 1972 that the Cambridge and Walthamstow breweries followed suit.

In 1977 Ellerman added Tolly Cobbold to its brewing interests (pipping Northern Foods, who had bought the Hull Brewery and built up a twelve per cent stake in Tolly). In 1983 the Barclay brothers bought Ellerman Holdings, including Tolly Cobbold's and Cameron's breweries and various travel and shipping interests. The proposed sale of Cameron's to Scottish & Newcastle in 1984 raised fears for the future of the Ipswich brewery as well, with rumours of bids from Greene King, Guinness and others, though Ellerman stated that, 'We intend holding on to Tolly. . . . Another brewery could bring Tolly more into the south-east market and help utilise spare capacity at Ipswich.' But none of the few remaining south-eastern breweries seems likely to fall Tolly's way in the forseeable future.

By the time of the Ellerman take-over, Tolly Cobbold's beer generally had a poor reputation, with the main criticisms relating to inconsistency and a thin texture ascribed to the use of adjuncts in the mash, including wheat flour and, so rumour had it, potatoes and even onions. The beers suffered further from being served under top pressure in most of the pubs. Ellerman deserve credit for changing all this, with the ordinary mild and bitter improved, the excellent Original bitter launched to replace Cantab, and the winter Old Strong ale retained. In addition, brewery redevelopment has been allocated substantial funds, and new pubs have been bought in areas previously devoid of Tolly tied houses – notably Norwich, Peterborough, Milton Keynes and London, where the Westmoreland Arms in Marylebone (ex-Watney's) was purchased in 1981. Conversely, pubs leased from Bass and Whitbread have reverted to the national brewers, and quite a number of low-barrelage country pubs have been disposed of. Eighteen properties regarded as surplus to requirements were disposed of in 1984 alone.

The traditional beer, now distinctive and very palatable, and served traditionally in 300 of the 340 pubs, consists of a dark and fairly sweet mild (1031); bitter (1034), a light and well-balanced drink which is the biggest seller; Original Bitter (1037), surprisingly only slightly stronger than the ordinary bitter but dry-hopped for extra sharpness; and Old Strong (1046), a very good and almost black winter ale. Tolly's bottled beers are extraordinarily popular, given the decline in this sector of the market nationally: even in the mid-1970s they still accounted for the majority of production. By far the best is '250', a very strong (1073) pale ale first brewed to celebrate Cobbold's 250th anniversary but kept in production by popular demand.

9 North-West England

There are still twenty-two established breweries in the North-West, though three of them (the new Bass brewery at Runcorn, Scottish & Newcastle's Royal Brewery in Moss Side, Manchester and Matthew Brown's Workington lager brewery) produce no traditional beer. Even so, this is a reduction of nearly two-thirds from the sixty or so which were brewing in 1960, with bankruptcy, lack of family successsion, and aggressive take-over bids all taking their toll.

The four Big Six producers of traditional beer are Allied at Warrington (Tetley Walker), Whitbread at Samlesbury (mostly a processed beer factory) and Chester's brewery in Salford, and Watney, at Wilson's brewery in Manchester. Manchester also has local beers from Boddingtons' (which also controls Oldham Brewery), Holt's, Lees' and Hyde's, with Robinson's in nearby Stockport. Warrington is the home of the biggest regional brewer, Greenall Whitley, and the Burtonwood Brewery is nearby. Boddingtons' third brewery, acquired in 1985, is the Liverpool brewery of Higson's, the last of Merseyside's independents. And both Matthew Brown and Thwaites brew in Blackburn.

Further north the picture is less rosy. Mitchell's keep alive the brewing tradition in Lancaster, but their neighbours Yates & Jackson were taken over by Thwaites in 1984, with the loss of their excellent beers. In Cumbria, Hartley's are still brewing but are now a subsidiary of Robinson's, and though Jennings is still independent there have been ominous signs of possible take-over in the 1980s. Cumbria has already lost a number of brewing concerns in the 1960s and 1970s, with the closure of Glasson's brewery in Penrith (bought by Dutton's in 1960) and Vaux's Kendal brewery, and the conversion of Workington Brewery into a lager-only factory after Matthew Brown acquired it in 1975. The biggest loss, though, was the State Management scheme in Carlisle, a nationalized brewing operation set up during the First World War which provided excellent and cheap beer but was wound up by the Conservative government in the early 1970s.

A label from pre-1972 State Management Scheme days

Theakston's (a Matthew Brown subsidiary) now run the brewery, but the SMS beers are no longer brewed.

Mild still accounts for a high proportion of sales in the North-West, and indeed many local brewers produced two milds until quite recently, but this practice is now in decline. Burtonwood and Greenall Whitley are amongst those to have discontinued their light milds, and other famous brands such as Chester's 'Fighting Mild' are now just memories. Bitter, especially pale-coloured bitter such as Boddingtons' and Matthew Brown's John Peel Special, is now the more popular choice.

BODDINGTONS'
Boddingtons' Breweries plc, Strangeways Brewery, Manchester.
Pubs: 280. Very substantial free trade, including London. Beers: Mild (1032), Bitter (1035).
Oldham Brewery, taken over in 1982, and *Higson's*, acquired in 1985, still brew separately and are therefore listed in their own right.

Boddingtons' were one of the stars of the real-ale firmament in the 1970s, renowned both for their beautifully sharp and very pale draught bitter and for their rejection of Allied Breweries' take-over bid in 1969–70. Today they are as likely to be reviled as revered – ironically because of their own re-emergence as an aggressive regional brewery (simply continuing a trend which saw them annex a number of their competitors up to the 1970s) rather than because their beers have suffered. It is particularly ironic because in the view of many connoisseurs (though not of the brewery) the beers *have* changed, the bitter in particular having become a good deal less distinctive.

Had Allied Breweries succeeded in 1969, almost two centuries of brewing independence would have been at an end. The first Strangeways Brewery was established by Caister & Fray in 1778, though neither partner lasted long, and by 1809 Hole & Porter (the former a member of the leading Newark brewing family) were the partners. In December 1832 they engaged Henry Boddington, then aged nineteen, as a clerk, and within a few weeks, following the discovery that a more senior clerk had been diverting some of the firm's money for his own benefit, he gained the first of several promotions which eventually saw him taken into partnership in 1848. Just four years later he became the sole proprietor when his partners withdrew, their confidence in the business apparently drained away by increasing taxes and oppressive legislation. The brewery itself was in disrepair, and the level of trading was at a very low ebb.

Henry Boddington's confidence was undiminished, however, and it was well founded, for by 1872 he had trebled sales and had even bought the small Bridge Brewery at Burton-on-Trent, together with pubs spread from south Cheshire to the Fylde. The Bridge Brewery was managed by Boddington's son Harry and concentrated on producing the pale ales for which Burton had become famous. By 1876 Boddingtons' was Manchester's biggest brewery, and Harry, who had originally confided that 'I like brewing very well, although there is a good deal of standing and walking about slowly', was assuming control of the business. Cash for expansion was generated when Boddingtons' became a public company in 1888, the family retaining a third of the shares.

This expansion involved the acquisition of a succession of other breweries, strengthening the network of tied houses. First was Hull's Brewery of Preston, with some sixty pubs, in 1900; then the Isle of Man Brewery became Boddingtons' Brewery (Isle of Man) Limited, with thirty-four pubs. The brewery was closed and became a depot, but in 1922 Boddingtons' withdrew from the island. The Burton venture had already ended, in 1912, largely because scientific advances meant that pale ale could just as easily be produced at Strangeways, using 'Burtonised' water. In 1940 the Strangeways brewery was destroyed by fire following a German bombing raid, a temporary setback but one which provided the opportunity to rebuild and re-equip the brewery, a task which was tackled with some determination.

The same determination saw off a number of would-be predators, from Walker's in the 1950s (who were told, 'First, your price is wrong. Secondly, I think we have a great future') to Allied in 1969. Beset by Allied and Whitbread, both of whom were now substantial shareholders, Boddingtons' grew in the 1960s in order to survive, taking over Richard Clarke of Stockport and sixty-five pubs in 1962, and J. G. Swales of Hulme, with thirty-eight houses, in 1971 (within twelve months trade in the Swales houses was seventy-nine per cent higher, a commentary on the relative popularity of the two companies' beers). By then Boddingtons' had seen off Allied (as has been described in Chapter 2), with the help of Whitbread and another of their largest shareholders, Britannic Assurance.

Since the historic victory Boddingtons' beer production has boomed, with the low-priced and full-flavoured bitter an increasingly familiar sight on northern bars in the 1970s and even further afield in the 1980s. Boddingtons' are adamant that the recipe of the beer has not changed since 1971, and during 1984 they conducted market research which they say 'confirmed the high standing of Boddingtons' locally brewed cask-conditioned bitter in the customer's eye'; others are less sure, feeling that it is darker, less bitter and less individual in character. Nevertheless, it is sold in traditional form in all the tied houses. The same cannot be said of the mild (1031), which is in a declining number of the pubs and could follow the same path as the once-popular winter Strong Ale, withdrawn in 1982.

The Boddington group, enlarged by the acquisition of Oldham Brewery in 1982 (at a premium of £8.7 million over the net asset value of £15 million: no wonder the Oldham directors recommended acceptance, and Allied and Whitbread quickly sold their ten per cent stakes in Oldham) and Higson's in 1985, now has about 530 pubs and, in addition to Boddingtons' cask beers, has keg beer from Oldham and a brand-new lager plant in Liverpool. Assuming that Whitbread's twenty-two per cent stake in the enlarged company is protective rather than threatening, future success seems assured. Perhaps the beer, too, will recover its standing amongst Manchester drinkers?

MATTHEW BROWN
Matthew Brown plc, Lion Brewery, Blackburn, Lancashire.
Pubs: 550. Beers: Lion Mild (1031), Lion Bitter (1036), John Peel Special (1040).
Also owns *Theakston's*, with breweries at Masham and Carlisle, and operates the Lakeland Lager Brewery at Workington.

Matthew Brown, now revered as the company which fought off a take-over bid from Scottish & Newcastle, had previously been criticized for their reluctance to make traditional beer available to more than a minority of their tied houses, for their conversion of Workington Brewery into a lager factory and for their take-over of Theakston's in 1984, with the consequence of obvious fears for the long-term future of at least one of the Theakston breweries. Despite their new status they still need to do more to persuade ordinary drinkers that the company's intentions are not to concentrate on Slalom lager, supermarket beer and bright beer for the clubs.

The Theakston take-over was itself a strange and complex affair (as described on page 214), with Matthew Brown eventually triumphing over the whisky distillers Grant's and adding Old Peculier to their beer portfolio. But within months Matthew Brown were themselves forced to respond to bid speculation with a statement that, 'It is the Board's strong conviction that the continued independence of Brown is in the best interests of shareholders, employees and customers.' Scottish & Newcastle were not so convinced, however, bidding £88 million in March 1985, £100 million later in the same month and, after an inept Monopolies and Mergers Commission report had sanctioned a take-over, £125 million in November 1985. S & N quickly lifted its shareholding to twenty-six per cent (seven per cent more than the combined stakes of the main Brown defendants, Whitbread Investment and Britannic Assurance) and when, after a titanic war of words, the bid closed at 3.30 p.m. on 11 December it had 47.5 per cent of the shares. Still greater drama unfolded as S & N's advisers announced victory at 5 p.m., claiming 50.3 per cent; but on the following day the Take-over Panel disallowed the shares gained after the bid closed, and delivered a rebuke on some of the methods used to try to persuade shareholders to accept the offer.

No doubt things were less frenetic when the original Matthew Brown was building up his business from its humble beginnings in a Preston beerhouse in 1830. His brewhouse in an adapted shop in Pole Street quickly proved popular, and he steadily built up the nucleus of a tied estate in the Preston area, buying, building and loan-tying pubs to ensure outlets for his beer. In 1875 the company was incorporated, with eighty per cent of the shares offered to the public and twenty per cent retained for his two sons, one of whom died soon afterwards, while the other was actually voted off the company's board in 1880. By his death in 1883 Matthew had recruited his son-in-law Joseph Smith to see the company, now with 110 pubs, into the twentieth century.

Major expansion began in 1913, when the Cunningham & Thwaites pubs were acquired on a 999-year lease, and continued in the early 1920s with a number of Preston firms eliminated. The Pole Street Brewery could hardly cope, and so Nuttall's Lion Brewery in Little Harwood, Blackburn, was bought, not just for its 200 pubs but also as the new base of Matthew Brown; all brewing was quickly concentrated at the larger Blackburn brewery. Meanwhile, another major initiative, that of expansion into Cumbria, had begun with the acquisition of Dalzell's brewery and its seventy-two pubs in the Whitehaven area. Later purchases (the Cleator Moor Brewery in 1947, Brockbank's in 1954 and the Workington Brewery, bought from Mount Charlotte Investments in 1975) have created a local monopoly of Matthew Brown houses in west Cumbria.

Traditional beers are available in only 200 or so of the 550 tied houses, and in Cumbria, with its local brewery capable only of churning out Slalom lager, the proportion is much lower – not much more than ten per cent. When they can be found, the beers are sufficiently distinctive to warrant sampling. Lion Mild (1031) is admittedly somewhat thin, as its gravity indicates, but it is a delicate, nutty-flavoured dark brew which still accounts for as great a proportion of production as Lion Bitter (1036), a malty and pleasant beer. Much the best pint, though, is John Peel Special (1040), introduced in 1980 partly to irrigate the Cumbrian beer desert. This is a very pale premium bitter with a delightfully dry, hoppy palate, and it deserves to be better known; with luck, it will get the chance to become so in many more of the pubs.

BURTONWOOD
Forshaws Burtonwood Brewery plc, Bold Lane, Burtonwood, Warrington, Cheshire.
Pubs: 290. Beers: Dark Mild (1032), Bitter (1036), JBA Premium (1039).

Burtonwood Brewery is bigger than most people realize, with a widespread network of pubs throughout the North-West and North Wales; it is still family-controlled, with the Dutton-Forshaws (who also have extensive interests as car-salesmen) and Gilchrists firmly at the helm. Following a recent boardroom reshuffle, it is highly ambitious, having failed with a £13 million bid for Border Breweries in 1984 only because they were opposed by the powerful combination of Marston's (the successful bidders, who immediately closed down the Wrexham brewery; Burtonwood had pledged to keep it open) and Whitbread, who owned a large slice of both Border and Marston's. One of Burtonwood's directors is on record as saying, 'If anyone would care to join us, we'd be glad to have them', and in 1985 the chairman described his company as 'still looking actively for acquisitions'.

Until recently extra trade might have been difficult to handle,

Two Burtonwood beer labels

since the brewhouse and some of the plant are relatively old and in need of attention. But a £3 million rights issue was launched in 1985 to pay for brewery redevelopment, including new kegging plant and an extension to the bottling stores. Burtonwood has also spent heavily on its tied estate in order to bring its 290 pubs, extending from Blackpool to northern Shropshire and from Anglesey to the Yorkshire borders (with just one pub in 'enemy' territory) up to modern standards.

The estate has been acquired gradually since the foundation of the firm in 1867, though there have also been a number of successful take-over bids. The founders were James and Jane Forshaw, who bought the farmland on which the brewery now stands largely because of the excellent supply of brewing-water. James Forshaw had previously been at the Bath Springs Brewery in Ormskirk; later Forshaw's Brewery Company was registered here, in 1898, but it survived only three years before being taken over by Ellis, Warde & Webster of the Snig's Foot Brewery in 1901. Ironically Ellis & Co moved to the Bath Springs Brewery, and by 1948 the Snig's Foot Brewery was back in Forshaw hands, when Burtonwood acquired Ellis's successors there, Richard Knowles & Sons. The Snig's Foot itself is now a fine, traditional Burtonwood pub.

The company has not always been free of problems (a number of the tied houses had to be sold to Higson's in 1925, and a further twenty-three pubs were sold to Tetley's in 1949) but it has been acquisitive, too, moving increasingly into North Wales from the 1930s onwards. Amongst the pickings were Lassell & Sharman of Caergwrle, near Wrexham, and their fifty-seven pubs in 1945, and Fox's Castle Hill Brewery at Ewloe, together with sixteen pubs, in 1949. The result is that Burtonwood is one of the best represented breweries in North and Mid Wales with, astonishingly, more than a third of their 290 houses in the Principality. Closer to home the small Standish brewery of J. B. Almond was bought in 1968; the brewery was closed and adapted to become the wine and spirits centre for the whole company, although this has now been relocated in Wigan. Despite all this growth, the Forshaws, Gilchrists and Almonds retain nearly half the company's shares, and Burtonwood (which has also moved into new fields, owning part of Haydock Park racecourse) looks secure against an unwanted take-over bid.

Regrettably, the two remaining long-established Burtonwood

beers have no great reputation. Light mild, a poor seller but highly regarded, has been withdrawn, and Dark Mild (1032), a creamy and smooth ale, seems less distinctive than its defunct counterpart. The bitter (1036) is well balanced and pleasant; a hoppier version was sold for a while as Almond's Bitter, but this has now been replaced by JBA Premium (1039), first brewed in 1986 and available only on handpump. Burtonwood's stronger premium bitter, Top Hat, is a 1046 brew which unfortunately is marketed only as a keg and bottled beer. The vast majority of Burtonwood pubs sell traditional beer, though it is worth noting that a few which do not have taken to using handpumps to dispense their keg or tank ale.

GREENALL WHITLEY

Greenall Whitley plc, Wilderspool Brewery, Warrington, Cheshire.
Pubs: over 1,100. Beers: Local Mild (1034), Local Bitter (1038), Original Bitter (1040).
Also owns *Davenport's*, Shrewsbury & Wem (*Wem Brewery*) and *Shipstone*, which are described separately.

It is almost an insult to call Greenall Whitley a brewery company, although beer still brings in about three-quarters of the company's profits. Expansion through acquisition of other breweries was the norm until the 1970s, but more recently expansion has included diversification of very considerable proportions. Separate divisions now run distilleries (making the famous 'wodka from Warrington' as well as gin), soft drinks, cider (Symond's of Herefordshire, where Greenall's took a controlling interest in 1984), hotels (including De Vere Hotels in Britain, and Treadway Inns, operating eight hotels in the United States), bingo halls and snooker clubs, amusement machines and other leisure interests.

The brewing business dates from 1762, when Thomas Greenall established his first brewery in St Helens. Twenty-four years later he formed a partnership to build the Saracen's Head Brewery on Wilderspool Causeway, Warrington, and on 10 January 1787 the first brew was mashed at this, Greenall's surviving north-western brewery. The Whitleys married into the Greenall family in 1824, and Greenall Whitley & Co was formally incorporated in 1880. Well before this, however, the Greenalls had begun to diversify, into textiles (where they were unsuccessful, despite the growth of the industry all around them), canals, railways, banking, water-

The mashing-stage at the Groves & Whitnall brewery in Regent Road, Salford in 1890

supply and glassmaking. Pilkington's world-famous glassworks in St Helens was known as Greenall & Pilkington in the 1830s, but the Greenall interest was later sold.

There were a number of acquisitions of breweries in Lancashire and Cheshire in the early part of the twentieth century, but Greenall's rise to regional prominence really began in 1949, with the acquisition of the Chester Northgate Brewery and its 140 pubs. Succeeding years saw the purchase of the Shrewsbury & Wem Brewery Company (1951), Magee, Marshall of Bolton (1958) and Groves & Whitnall, a major company based at the Regent Road Brewery in Salford (1961). The end of Magee, Marshall's independence was particularly sad, for it had been a very go-ahead brewery; founded in 1853 by David Magee at the Crown Brewery in Cricket Street, Bolton, the firm had acquired Marshall's One Horseshoe and Grapes breweries in 1885 and then took over Bell's Brewery in Burton-on-Trent in 1902, intending to brew pale ale there but never actually doing so, although water from the brewery well was transported all the way to Bolton.

The late 1960s saw the start of a series of closures of these subsidiaries, at Wellington (the Wrekin Brewery, bought in the

1950s) and Chester in 1969, Bolton in 1970 and Salford and St Helens – the original site – in 1972. Greenall's view is that, 'These were tough decisions to take, but proved to be correct. Concentrating brewing at the main Wilderspool brewery . . . keeps brewing costs down and improves efficiency.' Since then Shipstone's brewery has been acquired, in 1978, and together with the Wem brewery and Davenport's (a 1986 purchase), this is still in operation. The beer is blander, though, and the style of the company has changed, with the creation of what Greenall's call 'an attractive "80s" appearance, created through new signs and the latest "fun" pubs'. And 1985 saw Greenall's capture (and close down) the tiny Simpkiss brewery in the West Midlands, a tragedy described elsewhere.

Brewing still accounts for over fifty-five per cent of Greenall's turnover (and over eighty per cent of the group's profits in 1984) but there are only three traditional beers produced at Warrington, in an ultra-modern Steinecker brewhouse complete with lauter tun, wort kettle and whirlpool. Local Mild (the 'local' tag is ironic given the wide spread of pubs, from the Scottish borders to the West Midlands, and the company's record of brewery closures) is a dark 1034 brew with a malty taste, while Local Bitter (1038) is a bland, unmemorable ordinary bitter. Original Bitter (1040) is a slightly stronger beer, introduced in 1983, which again lacks a distinctive flavour. The beers are available in traditional form in a fair number of the pubs, but high-turnover pubs often sell processed beer stored in cellar tanks, whilst low-barrelage houses commonly use carbon dioxide to protect and dispense the beer.

HARTLEY'S
Hartley's (Ulverston) Ltd, The Old Brewery, Ulverston, Cumbria.
A subsidiary of *Frederic Robinson*, with 54 pubs. Beers: Mild (1031), Bitter (1031), XB (1040).

Controversy rages as to whether the 'Robinsonization' of Hartley's, following the agreed take-over in 1982, has been beneficial or potentially disastrous. On the one hand a number of basic, even run-down pubs in need of improvement have been altered and sometimes sympathetically extended, and the beers have been left alone, whilst their sales have increased; on the other the Robinson style has increasingly come to dominate the tied estate, and characterless one-roomed pubs have replaced some of the intimate, many-roomed Hartley's houses.

The take-over spelled the end of independence for a small company which had been founded in 1819 (on a site where brewing had already been taking place for sixty-four years) and bought by Robert and Peter Hartley, who came from Blackburn, in 1896. No strangers to the world of take-over, Hartley's bought the New Brewery in Ulverston, also in 1896, and then the tiny Kendal firm of Edward Hetherington, whose Gillingate Brewery served just three pubs, in 1918. Since 1932 Hartley's had brewed for the public houses owned by James Thompson of Barrow-in-Furness. Thompson's beer was apparently entirely separate from Hartley's and was brewed under the supervision of their own brewer. When Whitbread took over Thompson's in 1966, the arrangement persisted, with the beer supplied by Hartley's to the local Whitbread houses accounting for a quarter of the Ulverston brewery's output.

This extra business was crucial for Hartley's, whose own pubs took only seventy barrels a week between them in the 1950s. Energetic promotion of a new product, XB, was the main reason for a sixfold increase in output within twenty years. XB is now by far the biggest seller, with the ordinary bitter (which is not in many houses) and mild together making up less than a quarter of production. Free trade in particular, now accounting for some thirty-per-cent of sales, demands the XB in quantity, even outside the traditional trading area of south Lakeland.

But still doubts persist about the future of the brewery, which as recently as 1978 inspired the comment from a former managing director that, 'If Hartley's were taken over or shut down, there would be a riot all over Furness.' But no riot ensued when Robinson's won over the controlling families – the Robinson-Hartleys and the Darlings – four years later, with a pledge to retain the brewery for at least five years, and Robinson's themselves say that, 'The tradition [of brewing in Ulverston] is in safe hands.' It is quite a tradition, for the brewery is coal-fired and boasts an ancient cask-washing machine and open hopback, and the beer is racked solely into wooden casks.

Hartley's beers are a quite pleasant dark mild (1031) and an ordinary bitter (also 1031, but an entirely separate brew) which is crisp, light and extremely tasty for its low gravity, together with XB (1040), the premium bitter which has made Hartley's name for them. Malty and full-bodied, it is very drinkable and very easily recognizable. The bottling line closed in 1979, and there are no processed beers: the entire output not only leaves the brewery in wooden casks as traditional draught beer but is served as such, usually by handpump. Some of the pubs are outstanding examples of unspoilt Lakeland inns – such as the Golden Rule in Ambleside and the Outgate, a former tollhouse near Hawkshead, which brewed its own beer, known as BXB, until it became a Hartley's house.

HIGSON'S
Higsons Brewery plc, 127 Dale Street, Liverpool, Merseyside.
A subsidiary of *Boddingtons'*, with 160 pubs. Beers: Mild (1032), Bitter (1037).

Merseyside's last independent brewery company finally reached the end of the road in 1985, when it was swallowed up by Manchester's most acquisitive brewers, Boddingtons', in a £26 million deal which ironically secured the future of the brewery only because of its recently installed lager-producing capability. A decade or more of uninspiring results (the projected £1 million profit in 1985 was actually twenty-five per cent less than the corresponding 1975 figure), brought about by heavy investment in the brewery combined with Higson's dependence on trading in the recession-hit Liverpool area, eventually led the ruling Corlett family to the conclusion that there was no alternative to take-over.

The brewery had seen a number of ruling families since its foundation in 1780, when William Harvey established the business at 64 Dale Street. With his son Enoch he built the extensive Cheapside Brewery, one of the largest in Liverpool in 1800, but his nephew Robert disposed of the business to Thomas Howard, a fellow brewer, in 1845. It was Howard who employed Daniel Higson, then a book-keeper with another brewery, as his office manager; clearly Higson impressed his new boss, for when Howard died in 1865 he left the brewery to Higson. The company registered in 1888 bore the name Daniel Higson Ltd, but within thirty years the Higson family in turn had sold their interest in the firm. Before this, Higson's moved again, to the Windsor Brewery in Upper Parliament Street in 1912. But in 1918, as a result of boardroom squabbles and financial problems, the company was put up for sale and, despite interest from the major brewers, was bought for £118,500 by J. Sykes & Co, a Liverpool wine and spirits merchant.

Higson's advertising

The Sykes' solicitor, W. E. Corlett, negotiated the deal and became chairman in 1919, remaining in that position for forty-one years. In 1923 Higson's moved again, this time to the former Cain's Mersey Brewery in Stanhope Street, Toxteth, a superb Victorian brewery. Higson's still had their financial difficulties – in the early 1930s they would have gone under had not their maltsters provided eighteen months credit – but they were acquisitive too, buying Spragg's of Wallasey in 1919 to build up their Wirral estate, and Joseph Jones of Knotty Ash, with seventy pubs, in 1927 (closing the brewery in the following year). Finance was the reason for the long-standing link with Bass, who still held a twelve per cent shareholding until 1985, for Bass arranged a low-interest loan to Higson's in 1923 in return for Draught Bass (and later Carling lager) being sold in Higson's pubs. It was after the loan was repaid in the 1950s that the big brewer was invited to take a shareholding.

Profits were low in the 1970s while the new brewhouse was being built, and after it opened in 1982 there was strong criticism of the taste of the draught bitter, which was said to have lost its distinctive and locally popular flavour. But it was the launch of

Higson's lager in 1984 which precipitated the end of the company as a separate entity, for Boddingtons', whose own lager sales, at just eleven per cent, were barely more than a quarter of the national average, saw the opportunity to replace their bought-in Whitbread lager brands with their own lager, brewed in a brand-new brewhouse (which Higson's, ironically, had seen as a concrete expression of their faith in their own future). When the Corlett family, controlling over thirty per cent of the shares, accepted Boddingtons' offer, the success of the deal was virtually assured.

Higson's beers, which are generally good value for money, inspire fierce loyalty on Merseyside but indifference elsewhere. They are a rather bland dark mild (1032) and a bitter (1037) which is fairly well-hopped and certainly popular locally but which has no great distinguishing characteristics. Less than half the pubs serve the beer by traditional means, though the advent of Boddingtons' style of management may yet improve this situation.

HOLT'S

Joseph Holt plc, Derby Brewery, Empire Street, Cheetham, Manchester.
Pubs: 90. Has recently re-entered the free trade. Beers: Mild (1033), Bitter (1039).

Holt's appears to have leapt into the twentieth century in recent years, publicizing its history, chasing the free trade and increasing production of its quality ales. I doubt whether anyone calling in to their splendid wood-panelled offices and enquiring the whereabouts of their pubs would have to wait whilst their locations were sketched on a sheet of A4 – yet this was my experience in the early 1970s. As a result of the changes, Holt's classic bitter ale is more widely available, and reassuringly its future and that of the whole enterprise appears to be much more certain.

The origins of the firm have been traced back to 1849, when the newly married Joseph Holt, who had been working as a carter at Strangeways Brewery (then controlled by the Harrisons, now the home of Holt's close neighbours Boddingtons'), began brewing on his own account behind a pub in Oak Street, Strangeways. In 1855 Joseph Holt moved into the Ducie Bridge Brewery when the owners bought a bigger brewery in Hulme, and began to brew porter as well as beer. In 1860 he built the Derby Brewery on the present site, as well as acquiring his first tied house, the Apollo in Cheetham.

Joseph Holt died in 1886 and was succeeded by his only son, Edward, who presided over a period of hectic activity as Holt's participated in the rush to buy tied houses, which was at its height in the 1880s and 1890s, and also extended the brewery, adding a bottling stores, cooperage and stables for the dray horses. Altogether thirty-nine pubs were bought between 1886 and 1895, six of them in 1890 alone, yet in the next four years expansion was even more rapid, with forty-two houses acquired. Two breweries were taken over and quickly disposed of, the Bentley Brewery in Prestwich (bought at auction with one pub) and the Crumpsall Brewery in Cheetham Hill, with eight outlets.

The concern became a private limited company in 1922, and six years later, after the death of Edward Holt, his son (also Edward) invited his brother-in-law Harold Kershaw to join the board – hence explaining the strong Kershaw presence today: Peter Kershaw is chairman and, with his son Richard, joint managing director. In 1951 Holt's became a public company, although the Holt and Kershaw families retained a majority shareholding. That situation still obtains today, but the growth in other holdings precipitated take-over fears which were allayed only in 1985, when two possible predators sold their share stakes: Boddingtons' (9.8 per cent) decided that, 'Any possible long-term benefits of this investment were outweighed by the holding cost', and Robinson's (up to thirteen per cent at one time) sold their shares to a variety of institutions and individuals. Much smaller stakes are, however, held by Whitbread, Vaux, Bass and John Willie Lees.

Now Holt's are free to concentrate on brewing their outstanding beers, which were praised for their 'uniform excellence and superiority' as early as the 1880s but which by the 1960s were regarded as 'so often vinegary that they were avoided by practically an entire generation' – though beer sales still managed an increase during the period. Since then their former reputation has been totally restored, and there is excellent traditional beer in all ninety pubs. The bitter (1039) is widely regarded as one of Britain's truly great draught beers: very dry and bitter, pale straw-coloured and bursting with flavour, it shocks the tastebuds of those used to bland, ordinary bitters, but it is well worth the effort of familiarization. The dark mild (1033) is a fine drink,

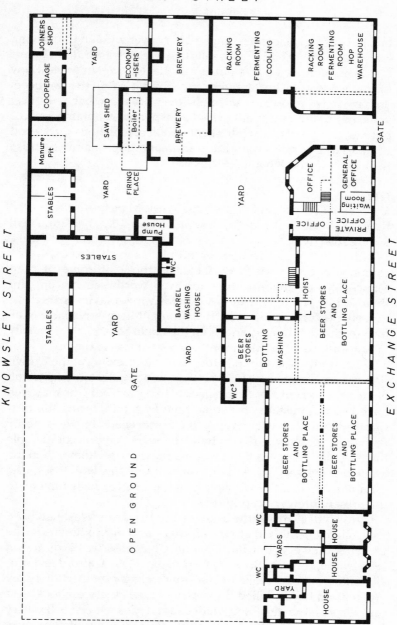

An architect's plan of the Derby Brewery in the 1930s

too, surprisingly clean-tasting and bitter yet with an underlying malty flavour; yet although it sells half as much as draught bitter, it is overshadowed by its illustrious stable companion and few from outside Manchester would recognize it. What the two beers do have in common are extremely low prices – probably the best value in Britain, as a result of low transport costs, high barrelage pubs (around half of all Holt's beer is delivered in hogsheads) and comparatively little expenditure on marketing, mainly because the beer is so good that it sells itself.

HYDES
Hydes Anvil Brewery Ltd, Moss Lane West, Manchester 15.
Pubs: 48. Beers: Mild (1032), Best Mild (1034), Bitter (1036), Anvil Strong Ale (1080).

Although Hydes' ales are by no means as well known, and certainly not as widely available, as those of some of their Manchester neighbours, they are well worth seeking out, for there is a surprisingly wide range of tasty low-gravity beers and a notably powerful winter ale. Real draught beer is available in all forty-eight tied houses, and even the Anvil Strong Ale is usually on sale during the winter months in two-thirds of them.

The founder was Alfred Hyde, who was brewing at the Crown Brewery in Audenshaw in 1863 (the company still has possession of a ledger with entries dating back to that year) but who may well have been in business for some time by then. During the next three decades Hydes were much-travelled brewers (only Batham's, perhaps, of surviving brewers have a comparable history), operating from the Victoria Brewery in Hulme, then the Mayfield Brewery in Ardwick and, from the late 1880s, the Monmouth Street Brewery in Rusholme before finally moving to the present brewery in July 1899.

Until Hydes' arrival the brewery in Moss Lane West had been controlled by the Greatorex brothers and was known as the Queen's Brewery, but the Greatorex business was acquired in 1898 by the Empress Brewery Company of Old Trafford (eventually taken over by Walker Cain, themselves now part of Tetley Walker and hence Allied Breweries). Hydes therefore took over the surplus brewery and traded as Hydes Queen's Brewery Limited until 1944, when the present name was adopted because it was felt to be more in keeping with the company's trademark.

Hydes is still wholly family-controlled, with the great-grandsons of the founder now acting as joint managing directors, and virtually all the voting shares organized into seven holdings, including family trusts, in which the Hydes have a controlling interest. The continuing independence of the company is an important objective: the chairman is on record as saying that, 'We have no intention of being swallowed up by any of our larger brethren.'

The forty-eight pubs are located mainly in the southern suburbs of Manchester, though there are one or two in or near to the city centre, such as the tiny but comfortable Grey Horse in Portland Street and the incomparable Jolly Angler in the back streets near Piccadilly (not one to take more sensitive souls to, this!). One or two lie further afield in Cheshire – the Bull's Head in Lymm is a good example – including the company's newest pub, the Hoop & Mallet in Warrington, which was opened for Christmas 1985. There is also, inexplicably, a clutch of pubs in the urban villages around Wrexham.

Hydes, now the only brewery in Manchester to produce two separate milds, offer a choice of mild (1032), a darkish and fairly sweet brew, and Best Mild (1034), much lighter in colour and with a subtle, pleasantly hoppy flavour. This is a fine low-gravity beer, yet I have been actively dissuaded from drinking it by a Hydes' landlord. Despite this the milds between them account for something approaching half of production. The bitter (1036) is a well-balanced if not especially memorable brew, and the beer list is completed by Anvil Strong Ale (1080), a rich and heavy draught barley wine which can be found between November and March. A keg bitter and Amboss draught lager are also produced.

JENNINGS
Jennings Brothers plc, Castle Brewery, Cockermouth, Cumbria.
Pubs: 76. Beers: Mild (1034), Bitter (1034), Marathon (1041).

Jennings' moment of glory was in 1973, when the company fought off a take-over bid from Mount Charlotte Investments, part of the Grand Metropolitan group, which had only recently acquired Watney's and Truman's, and therefore presumably regarded Jennings and their fellow Cumbrians at Workington Brewery, for whom they made a (successful) simultaneous bid, as easy meat. It was not to be, for despite raising their offer several times Mount Charlotte could tempt only a minority of the shareholders, and eventually they were forced to admit defeat. It was a

Jennings Brewery and Cockermouth Castle

victory for local independence: as Jennings themselves say, 'The shareholders were clearly reluctant to see destroyed something that had become a part of local life.'

Unfortunately not everyone is convinced by social rather than financial considerations, and rumours have continued to link Jennings with other brewers. Sometimes the threat has been more tangible. In 1981 Matthew Brown built up a thirteen per cent shareholding, and four years later Samuel Smith's suddenly increased their seven per cent stake to eighteen per cent apparently because Jennings switched their allegiance from Smith's Ayingerbrau lager to Tennent's (part of a wider trading agreement with Bass). Peace was declared when Ayingerbrau found its way back onto Jennings' bars, but somewhat worryingly Smith's retain their shareholding.

Jennings' own activities only confuse the issue. On the one hand they have invested substantially in expanding the business, buying Underwood's, a Maryport firm of mineral-water manufacturers, in 1984 (and a similar though smaller Dumfriesshire company in 1985), and opening a brand-new brewhouse in 1985. On the other hand their vacillation over the Tennent's lager contract cost them their trading agreement with Bass, which included the lease of four pubs in south Cumbria, and although they signed a replacement trading agreement with Tetley's, no leased pubs were included. Furthermore, Jennings' own tied estate has contracted sharply in recent years, from ninety to seventy-six, as a quite severe pruning of uneconomic houses has been put into effect – including their two Newcastle pubs, which Jennings ran for only two or three years.

This contraction contrasts with the steady expansion of the business from its origins in a home-brew house in the nearby village of Lorton in 1828. The firm, initially run by the four Jennings brothers, moved to Cockermouth in 1887, when the volume of trade outstripped the capacity of the Lorton premises; the choice of the new site, immediately below Cockermouth Castle and just above the confluence of two rivers, the Cocker and Derwent, was based on the ample supplies of pure well water in the area. Capital for the new venture was raised by incorporating Jennings Brothers as a public company in 1887, though by now the Jennings family was no longer involved. Jennings' major acquisition was Faulder's Browfoot Brewery in Keswick, bought in the 1930s after it had ceased brewing in 1926 and bought in supplies of Jennings beers for its pubs. In 1972 eight more pubs were acquired as a result of the winding-up of the State Management Scheme in Carlisle.

The three traditional Jennings beers are mild (1034), a medium-dark, malty and quite full-flavoured brew, bitter (1034), surprisingly dark and malty but nevertheless bitter-tasting and consequently a really individual beer, and Marathon (1041), a smooth malty premium bitter introduced in 1985. With luck Marathon should last longer than two other brews Jennings have tried in the 1980s, the premium-strength Special bitter, which failed to generate much interest, and the much weaker Cumbria Pale Ale, a bland light mild which lasted only a matter of months. Jennings also brew a keg bitter, mainly for the free trade, which overall accounts for more than a quarter of output.

LEES

J. W. Lees & Co. (Brewers) Ltd, Greengate Brewery, Middleton Junction, Manchester.
Pubs: 127. Increasing free trade. Beers: GB Mild (1032), Bitter (1038), Moonraker Strong Ale (1074).

John Willie Lees is an excellent example of a long-established family brewery company which has survived almost in isolation, guarding its independence jealously and at the same time having no history of extravagant take-over bids for other similar businesses. The business was started in 1828 when John Lees, a retired cotton-manufacturer, bought some cottages and adjoining land in Middleton Junction and began to brew porter and ale. The venture was highly successful, and the founder's grandson, John Willie Lees, built a much larger brewhouse in 1888 – the nucleus of the present, much extended Greengate Brewery.

Disaster very nearly struck twice thereafter, however – first when John Willie Lees died childless in 1907 and the brewery was left in the hands of trustees, who allowed the business to stagnate with virtually no investment, and secondly when a nearby brewery tried to take Lees over in the 1950s, fortunately without success. Take-over now would be more expensive for a predator – Lees have seen beer production quadruple in the last twenty-five years and now have a tied estate of 127 pubs, a quarter of them in North Wales, the rest in north Manchester (all but eight of them serve real beer), together with night clubs in Burnley and Chester, a 'real English pub' in the French Alps, a wine shop and a major stake in Oldham Athletic Football Club. Any would-be bidder would also find that the Lees-Jones brothers, Richard and Christopher, fifth-generation descendants who are the joint managing directors, control all the company's shares through a series of family trusts.

The French connection is not so much a diversification as an accidental diversion, a Lees director having been asked to provide an English pub in the French skiing resort of Flaine. The result, the White Grouse, was built by a firm from Oldham and stocks 'bright' Lees bitter and Edelbrau lager as well as darts and shoveha'penny. The link with Oldham Athletic is more complex – a Lees director is also chairman of the football club, and there is a £¼ million sponsorship deal and loan to relieve the financial pressures on the club.

The buoyant sales of Lees traditional beers, and their growing presence in the Manchester free trade, may surprise those who remember the beers' less than glowing reputation some years back; indeed, there are those who find the present brews, which are still mostly racked into wooden casks, rather lacking in character. The most interesting, perhaps, is GB Mild (1032), a pleasant, darkish amber mild named after the initials of the Greengate Brewery. It was first brewed as recently as 1982, when it replaced both a light mild which was in only a few pubs and Best Mild, a dark, sweet brew especially popular in North Wales. Another relative newcomer is Moonraker Strong Ale (1074), a heavy and malty brew which is available during the winter months at selected pubs only. The bitter (1038), a creamy, light-coloured beer with a pronounced malty bite, is the mainstay, accounting for about half of total production.

Lees have followed a policy of expansion through building new pubs since the 1960s, with more than twenty new developments to date, and they have also expanded the brewery substantially, with new fermenting vessels and, in 1985, a new mash tun and copper to increase capacity by fifty per cent. There is even a long-term plan to rebuild on an adjacent site if necessary. As Richard Lees-Jones said in 1985, 'The popularity of good traditional ale has never been greater, and we have had to expand our capacity to keep pace.' It is to be hoped that another sentiment of his – 'We are determined not to be taken over' – deters the predators long enough to allow Lees to reap the benefit of this expansion.

MITCHELL'S
Mitchell's of Lancaster (Brewers) Ltd, 11 Moor Lane, Lancaster.
Pubs: 50. Beers: Mild (1034), Bitter (1036), Special Bitter or ESB (1045).

The sad demise in 1985 of Yates & Jackson leaves Mitchell's as the only surviving brewery company in Lancaster. Happily, the company seems set to survive for some time to come, having moved from their own very cramped city-centre site to Yates & Jackson's brewery (declared redundant by its new owners) in March 1985, and having begun to expand their tied estate after years of apparent inertia. Their beers perhaps lack the dry subtlety of Yates & Jackson's much-missed bitter, but now that they have the use of the natural well water in Brewery Lane (Mitchell's previously had to rely on town water, which could be far from ideal for brewing in the summer months), a further enhancement to the reputation of their already fine ales is possible.

Mitchell's celebrated their centenary in 1980 (with a magnificent, strong Centenary Ale (1080) which was occasionally available as a draught beer but, though it is still produced, is now bottled only), but William Mitchell had actually been engaged in brewing for some nine years prior to the completion in 1880 of his brand-new brewery in the centre of Lancaster. He leased the Black Horse in Common Garden Street in May 1871, sharing a brewhouse with the adjacent Bull's Nut public house; three years later he bought the New Inn in Market Street (still a Mitchell's house), supplying it from the Black Horse brewery. With a loan of £50,000 from a friend, Andrew Rhodes, he built his own brewery behind the New Inn and began to look for trade, using a good deal of imagination in his search. While the pipeline taking Thirlmere's water to Manchester was being laid, he bought up houses alongside it, such as the Redwell Inn at Over Kellet; then, once pipelaying in the area had finished and trade had returned to normal, he sold them again.

Mitchell recognized the restrictive nature of his site (the original entrance, by the side of the New Inn, was so narrow that his horse-drawn drays had to back down the yard) but his attempt to move to a site on Moor Lane, which he had bought in 1915, proved unsuccessful. His application for building consent was approved subject to the building being dressed in stone, and since he was unable to afford the extra expense, Mitchell had to soldier on using the original brewery. When William Mitchell died in 1919, the brewery controlled about a hundred pubs, though many were heavily mortgaged and his executors – running the brewery for his daughter Annie – saw several closed under the 1913 Licensing Act and others sold, including the Royal

An advertisement for Mitchell's beers

King's Arms Hotel in Lancaster, as John Barker, who had married Annie Mitchell and acted as brewery manager before William Mitchell's death, struggled to keep the business afloat in the 1920s.

The opening of a wine and spirits department (bottling its own brands of whisky and pure pot still rum) and the management of William Mitchell Barker senior – in charge from 1936 to 1970 – ensured Mitchell's survival, and in 1965 the executors of William Mitchell finally gave way to the present company. Today the third generation of the Barker family run the firm, which has fifty pubs (an increase of three in the 1980s) and in 1985 gained a bigger site at last by buying Yates & Jackson's Brewery Lane premises from Thwaites of Blackburn.

Three traditional beers are produced by Mitchell's, and all of them are quality products well worth sampling. Best Dark Mild (1034) is a smooth, creamy brew once known as Country Mild because of its popularity in the more rural pubs, and Bitter (1036), made with Yorkshire pale-ale malt together with some maize and crystal malt, is a nutty, malty beer. The premium beer is ESB or, more commonly, Special Bitter (1045), introduced in May 1972 and widely respected as a very full-bodied but also well-hopped bitter ale. Exceptional pubs in which to try these beers include the Carpenters Arms in Lancaster, dating from 1300 and one of the oldest pubs in England, and the Old Hall in Heysham, formerly an Elizabethan manor house.

OLDHAM
Oldham Brewery Co Ltd, Albion Brewery, Coldhurst Street, Oldham, Manchester.
A subsidiary of *Boddingtons'*, with 87 pubs. Beers: Mild (1031), Bitter (1037).

A brewery with an interesting history, Oldham faces an uncertain future. The take-over by Boddingtons' in 1982 was followed by assurances that the brewery would remain open for at least five years. This was uncharitably assumed to mean 'five years and a day', and since Boddingtons' have since expanded still further with the purchase of Higson's of Liverpool, with its more modern plant geared to lager production, the Albion Brewery may indeed now suffer once Boddingtons' are forced to the conclusion that three breweries in such a small area are one too many.

The beers which would suffer command a substantial local

Oldham Brewery's offices, constructed around 1910

following but are largely unknown outside the locality. Draught Mild (1031) is a dark, malty and rather sweet brew of no great distinctiveness; Bitter (1037) is more immediately recognizable, a light-coloured ale with an unusual taste – an acquired taste, in fact. The beers are available in traditional form, sometimes from wooden casks, in only a third of the eighty-seven pubs, alongside Boddingtons' beers in some instances, and the parent company seems to have put more effort into promoting Oldham's keg products.

The brewery deserves better than this, though. The company was formed in 1873 by a group of local businessmen specifically to purchase Boothby's Brewery in Coldhurst. Converted from an old hat factory in 1868, this brewery supplied five tied houses and some free trade. Business boomed in the 1870s (in 1875 alone thirty-two new cotton mills opened in Oldham, an indication of the excellent economic conditions in which the local breweries were operating), and by 1880 there were thirteen tied pubs. The Edge Lane Brewery was taken over in 1895, purely for its tied houses (the brewery was sold), and by 1920 75,000 barrels a year were being brewed – an output not exceeded until the 1970s, partly because of the inter-war recession (which saw production slump to only 40,000 barrels in 1933). One ironic bright spot in 1940 was the opportunity to supply some of Boddingtons' houses for a matter of months, when the Strangeways Brewery was destroyed during an air raid.

A steady decline in tied houses is evident from the 1940s onwards, perhaps indicating a lack of ambition. More than thirty Oldham pubs have been lost to redevelopment schemes since then, and in addition to these losses another five town-centre pubs in Oldham were sold for shopping developments in the 1970s. The bottling plant, too, was shut down in 1979. Perhaps because of this apparently passive attitude to development, Oldham became the subject of bid rumours in the 1970s (Northern Foods, then the owner of Hull Brewery, was reputed to have made an unsuccessful offer), and Boddingtons' bid, in 1982, was quickly accepted. Despite the optimism in the official history of the firm, published by Boddingtons' in 1985, that, 'You can still get a pint of O.B. beer all over Oldham. There's no reason to believe that you ever won't', there are inevitably doubts about the future of the brewery and its ales, which should be sampled while they are still available.

ROBINSON'S
Frederick Robinson Ltd, Unicorn Brewery, Stockport, Cheshire.
Pubs: 316. Beers: Best Mild (1032), Bitter (1035), Best Bitter (1041), Old Tom (1080).
Also owns *Hartley's*, taken over in 1982 but still brewing separately.

The Robinson family have been brewing in Stockport since 1865, but the foundations were laid twenty-seven years earlier when William Robinson, a corn and flour merchant from near Macclesfield, bought the Unicorn Inn in Lower Hillgate, Stockport, from Samuel Hole. He ran the pub as a retail business only and refused to open on Sundays, saying that, 'Six days is enough for any man to work', but when his younger son Frederic joined the business in 1865, the pub began to brew its own beer. The first outside customer was a Mrs Lamb, who ran the Bridge Inn in nearby Chestergate, and the first pub to be bought, on 3 May 1876, was the Railway Inn in Marple Bridge (still a Robinson's house but rebuilt in the 1930s and now called the Royal Scot).

When Frederic's son William joined him in 1878, the business consisted of two pubs and a horse and dray; twelve years later, on Frederic's death, there were twelve tied houses, and by 1920, when a private limited company was formed by Emma Robinson (Frederic's widow), together with William and his sons Frederic, John and Cecil, a secure financial basis had been laid for a series of bold moves designed to enlarge the business. First, two take-overs in the 1920s widened the scope of the firm: Schofield's Portland Brewery in Ashton-under-Lyne, with forty-two pubs, and Kay's Atlas Brewery, which brought with it another eighty-six pubs. The Atlas Brewery was closed in 1936, by which time a new brewhouse had been built at the Unicorn Brewery, but the 1930s was notable as the decade in which, against the national trend, Robinson's continued to buy pubs, especially in Cheshire and the North Wales countryside. The legacy of this policy is a whole series of attractive country pubs selling Robinson's beer today.

The acquisition of Bell & Company of the Hempshaw Brook Brewery in Stockport in April 1949, with 160 pubs, virtually doubled the size of the company and eventually made expansion away from the town-centre Unicorn Brewery site inevitable (a move hastened by the realization that the old brewery site was slowly subsiding). Hence since 1973 kegging, bottling and ware-housing have gradually been moved to a greenfield site at Bred-bury, four miles away, on land acquired in 1919. Brewing remains at the Unicorn Brewery, which can produce up to 125,000 barrels a year, for the time being. Finally, 1982 saw Robinson's expand further north with the take-over of Hartley's of Ulverston, a family firm owning fifty-six pubs; Hartley's brewery remains open at present, though since Robinson's already send beer a similar distance, to Anglesey, they may eventually decide to concentrate all brewing at Stockport.

Traditional beer-enthusiasts would, in that event, quite rightly mourn the loss of Hartley's beers but would at least be able to drown their sorrows with Robinson's very good range of real beers, which between them account for four-fifths of production. Much of Robinson's beer is still racked into wooden casks (a cooper's shop is still maintained) and a little of it is delivered to nearby pubs by horse-drawn dray. For its size Robinson's is, in fact, a commendably traditional brewery, and it is also a very private one, with the Robinson family still in control of the share capital and providing all the directors.

The Best Mild (1032) is an outstanding example of its type, quite pale-coloured and malty in most outlets but also produced with extra caramel to darken it for a few pubs, such as the excellent Old Pack Horse in Chapel-en-le-Frith. The ordinary bitter (1035) is a thin, surprisingly bland brew which is in only a minority of the pubs, but Best Bitter (1041) is a magnificent beer, very pale but full-flavoured and well-hopped. Finally there is the excellent Old Tom (1080), primarily a bottled barley wine but widely available in draught form during the winter months, as a very strong and not too sweet darkish winter ale.

THWAITES
Daniel Thwaites plc, Star Brewery, Blackburn, Lancashire.
Pubs: 411. Beers: Mild (1032), Best Mild (1034), Bitter (1036), Daniel's Hammer (1075).

Thwaites is the Jekyll and Hyde of independent breweries, with on the one hand a deeply traditional image fostered by outstanding 'mature draught' beers (the Best Mild collected the Beer of the Year award at the Great British Beer Festival in 1980) served traditionally in almost all the pubs, together with a series of superbly renovated traditional pubs and the dray horses which now make local deliveries, and on the other hand an uncompromisingly modern tower block of a brewery, with its ten-foot-high neon sign and very up-to-date canning, kegging and bottling facilities. The traditional image was also spoilt somewhat in 1984 when, in their first foray into the take-over market since 1956, Thwaites bought up and immediately closed down Yates & Jackson of Lancaster, another splendidly traditional firm whose marvellous ales are now just a memory.

The company's origins are highly unusual. Daniel Thwaites was an excise officer who turned to brewing in 1807 when he decided to buy one of the breweries he visited, in Eanam near Blackburn. The partnership of Messrs Counsell & Duckworth was therefore succeeded by Daniel Thwaites, and the brewery has remained under the control of his descendants until the present day. One of the few vestiges of the original brewery (or even of its Victorian successor) is the well-house, for although the well is no longer in use it remains as an indication of the reason for Daniel Thwaites' original interest, namely the natural water which was the secret of prosperity for this and other Blackburn breweries such as Matthew Brown, who now form the main local competition, and Dutton's, long since taken over and closed down by Whitbread.

Thwaites have risen to their present position as a major northwestern brewery, with more than 400 pubs, through a mixture of steady, unspectacular growth and sudden bursts of take-over activity. The most notable acquisitions have been Henry Shaw's Salford New Brewery, a Blackburn rival, taken over in 1923 with eighty-seven licensed houses; the Bury Brewery Company, adding about eighty pubs in 1949; the Preston Breweries in 1956; and, of course, Yates & Jackson in 1984. This latest purchase added forty pubs, tightly grouped in the Lancaster area, for some £5 million. Yates & Jackson's profits record had been unspectacular for some time, prompting them to say that, whilst the company was 'not in financial difficulties at the moment, we can't go on trading like this over the next decade, so we decided to go now when we could choose a partner who was able to give beer of the same quality'. Thwaites recouped some of their investment by selling Y & J's brewery to Mitchell's of Lancaster.

The newly acquired pubs were served from Thwaites' modern brewhouse, completely rebuilt between 1966 and 1972 at a cost of some £5½ million. Capable of brewing 300,000 barrels a year, the plant operates at about eighty per cent of capacity, producing three draught beers (also available as bright beers), a keg mild, the bottom-fermented Stein lager and a range of bottled beers, some for export to Spain and the United States. Together with this heavy investment in brewing Thwaites have also diversified, moving into agriculture with Yerburgh Farms in 1977 and three years later forming Shire Inns to run their hotel and catering interests.

The traditional side of Thwaites' operation is exemplified by their return to horse-power for local deliveries. At one time Thwaites stabled forty horses at the brewery, but by the 1950s they had been phased out, replaced first by steam wagons and then by motor vehicles. But in 1966 the horses were back, partly for publicity reasons (their shirehorses are now a familiar sight at fairs and carnivals) but also as the most economical way of moving barrels of beer around Blackburn. About half of Thwaites houses are within just a few miles of the brewery, with heavy concentrations in the Blackburn and Accrington areas, but there are notable exceptions, in the Potteries, Nottingham, Barrow and, since 1984, one pub in London. The pubs are generally refurbished to a high standard, and the brewery's sympathetic approach to alterations and extensions has won them national awards.

The beers sold in the pubs have also won a string of awards. Although the proportion of mild brewed is now below one-fifth of total output, a complete reversal of the situation in the 1950s, when mild made up three-quarters of sales, Thwaites have kept faith with two varieties, namely Draught Mild (1032), a dark nutty-flavoured brew, and Best Mild (1034), a lighter, malty beer which is held in high regard locally. Bitter (1036), creamy and well-hopped, accounts for more than half total sales. And in winter Daniel's Hammer (1075) is a draught barley wine available in about a quarter of the pubs, usually direct from the cask.

10 North-East England

The North-East is the only region in the country where the Big Six still operate more breweries than the remaining regional and local brewers. All but one of the remaining nine Big Six breweries produce traditional draught beer, the exception being the small Hope & Anchor Brewery in Sheffield, now used by Bass as an experimental brewery. Bass also run Stones' Cannon Brewery in Sheffield and the Tadcaster Tower Brewery. Whitbread have Tennant's Queen's Brewery in Sheffield and the Castle Eden Brewery, where the best of their northern beers are produced. Good beer is also produced at Allied's Tetley brewery in Leeds and at Webster's Halifax brewery, part of Watney since 1972. Two notable conversions to real beer are Scottish & Newcastle's Tyne Brewery in Gallowgate, Newcastle, where real draught Exhibition bitter is once again brewed, and John Smith's Tadcaster Brewery, part of the Courage empire, at last producing traditional beer again after a decade of producing bright beers only.

The 'independent' sector is thin on the ground, with three of the seven breweries still in operation owned by other brewery companies and a fourth, Cameron's, a subsidiary of Ellerman Holdings. Vaux operate three north-eastern breweries, at Sunderland, Sheffield (Ward's, taken over in 1973) and Thorne (Darley's, 1978). Theakston's, famous for its Old Peculier strong ale, was bought by Matthew Brown in 1984. But the two West Yorkshire breweries, Samuel Smith and Timothy Taylor's, share an intense conservatism and independence (and brew some good beers, too). Cameron's Strongarm and Theakston's XB are amongst the beers to seek out.

Three other breweries should also be mentioned, since in some ways they might have qualified for inclusion in this text. The Northern Clubs Federation Brewery in Newcastle is the larger of the remaining club breweries in the country and has been in business since 1919 – but it brews no traditional beer. Two old firms have returned to brewing since the 1970s, but the long break in production in each case rules them out as 'traditional' firms. The first is the Selby Brewery, which was founded in 1870 but ceased brewing for economic reasons in 1954 and sold its pubs to Dutton's of Blackburn (later taken over by Whitbread) ten years later. After a period as beer-bottlers, the firm began to brew again in 1972 and sells Best Bitter (1039) in one tied house and a few free trade outlets. The second firm is Clark's of Wakefield, whose Westgate Brewery was in irregular production, supplying the free trade, until the 1960s and was then resuscitated in 1982, supplying a tied house next to the brewery and some free houses. The beer in this case is Clark's Traditional Bitter (1038).

CAMERON'S
J. W. Cameron & Co Ltd, Lion Brewery, Hartlepool, Cleveland.
A subsidiary of *Ellerman Holdings*, with 480 pubs. Beers: Lion Bitter (1036), Strongarm (1040).

The Lion Brewery has had a chequered history, especially in recent years, but at present it roars on in semi-independent fashion, brewing a good deal of traditional beer (certainly by the standards of north-east England) but criticized by many for its policy on pub closures. The brewery dates from 1852 and was leased to John Cameron twenty years later; by the end of the

A Cameron's beer advertisement

nineteenth century local competition had been all but eliminated (Cameron's bought the two Church Street breweries, Rickinson & Sons and Nixey, Colclough & Baxter, in 1895) and the company began to spread its influence further afield.

Major expansion was, however, reserved for 1954, when Hunt's of York and the Scarborough & Whitby Breweries were acquired, netting 220 pubs in all. The West Auckland Brewery in 1959 brought in another eighty pubs, and when Russell & Wrangham of Malton was bought in 1961 from its previous owners, Melbourne's of Leeds, a further ninety pubs were collected. The enlarged Cameron's now had 750 pubs east of the Pennines, stretching from the Scottish Borders to Sheffield. But it had also acquired two outside shareholders, Ellerman Lines and Bass, and in 1974 Ellerman took over the brewery. Immediately there was a spate of pub closures – a hundred disposals in the first two years of Ellerman rule, and this issue has surfaced again on a number of occasions.

The sale of Ellerman Holdings (including Cameron's and Tolly Cobbold) to the Barclay brothers in 1983 fuelled fears that Cameron's might be at risk – fears which almost proved justified, for the sale of the brewery to Scottish & Newcastle for £44 million was agreed the following year. S & N pledged that Cameron's would retain its identity, but the closeness of the Hartlepool and Newcastle breweries made eventual rationalization highly probable, leaving S & N to supply their fifty per cent of the North-East beer market from their Newcastle plant. Happily the bid lapsed when it was referred to the Monopolies & Mergers Commission, and later in 1984 the Barclays decided to retain Cameron's and improve the profitability of the business themselves. Since then sixty-five off-licences have been sold to Vaux, pubs have been swapped with S & N, and pub sales have continued, leaving a leaner and fitter business.

The most encouraging aspect of Cameron's strategy is the increased reliance on traditional beer from handpumps, which Cameron's has found increasingly popular and profitable. Forty per cent of the 750 pubs sold cask-conditioned beer in the 1970s, and the proportion has now increased to sixty per cent of the slimmed-down tied estate, with the hope of at least eighty per cent in the near future. Two traditional beers are brewed these days, three more (Scotch, ordinary bitter, and mild) having been withdrawn around 1970. The weaker of the two survivors (a descendant of a former Russells & Wrangham brew) has been known since 1983 as Lion Bitter (1036), though it was formerly called Best Bitter (a name now reserved for the bright version of the same beer). A light, well-hopped bitter, Lion Bitter is available in over 200 tied houses, an appreciative free trade and, further afield, many Everard's houses as a result of a beer swap arranged in 1985. Everard's Old Original is available as a strong premium bitter in Cameron's pubs – a particularly welcome move following the reduction in strength of Cameron's own premium bitter, Strongarm.

Strongarm (now 1040) was launched as such in 1955, though it was first brewed in 1953 as Coronation Ale. The first Strongarm had an original gravity of 1047 and was especially popular on Teesside; by the 1970s the gravity had been reduced to 1042, and in a surprise move in 1983 a further weakening of two degrees was implemented. But the beer, which outsells Lion Bitter in pubs close to the brewery but is less likely to do so in far-flung parts of the tied estate, is still well balanced and retains much of its rich, satisfying and unusually malty flavour.

DARLEY'S
W. M. *Darley Ltd*, The Brewery, King Street, Thorne, South Yorkshire.
A subsidiary of *Vaux*, with 61 pubs. Beers: Dark Mild (1033), Thorne Best Bitter (1038).

Darley's proud and very private existence as a separate entity ended in 1978, when the family-controlled company sold out to Vaux of Sunderland. Yet, surprisingly, the firm does retain a separate identity, brewing its own beer for the sixty-one pubs and also for a certain number of pubs trading under the Vaux banner. Traditional beer from Darley's was fast disappearing before the take-over, with many pubs converted to bright, pressure-dispensed ale, ostensibly because of problems over beer quality, and Vaux deserve praise for the way in which they reversed this policy and emphasized the highly individual – some would say peculiar! – qualities of Thorne ale.

Darley's connection with brewing in Thorne dates from the middle of the nineteenth century, although there had been a brew-house on the site for a century before that. The founder, W. M. Darley, born in 1837, took over the present brewhouse, which dates from 1830 and still boasts fifteen slate Yorkshire squares (most of them late nineteenth century) in its fermenting room, together with a steam-jacketed mash tun installed in the 1920s. His grandson, Thomas, took over as chairman in 1929 and remained at the helm for fifty-three years until his death in 1982 at the age of ninety. It seems that it was Thomas Darley, the last direct descendant of the founder, who suggested the take-over to Vaux, a company with which Darley's already had trading links, and that it was the coming end of the Darley dynasty which prompted him to do so. The £3 million take-over was arranged because Tom Darley believed that Vaux would continue to operate the brewery, whereas other bidders, perhaps prepared to pay more, would close it immediately.

Vaux have kept faith with the brewery, though not all its operations have survived. The bottling line was an early casualty, though since throughput was down to fifty-four gallons a week and the machinery was antique, it might not have lasted long in any case. The end of Darley's bottled beers did however also spell doom for a cask ale, light mild, also bottled as Amber, but light mild was supplied to just one outlet, so that its future looked bleak anyway. A succession of changes have hit dark mild, which was also threatened by falling sales. Early in 1984 it was replaced by a very dark, lightly hopped porter-type brew, Chairman's Dark Ale, advertised as 'more than mild, better than bitter', which was brewed at Thorne and also supplied to Ward's pubs (also owned by Vaux). But by November 1984 the new beer had failed and was succeeded in turn by Dark Mild (1033), not dissimilar to the original Darley's mild ale.

The mild is, in truth, a little bland, whereas Thorne Best Bitter (1038) is a very distinctive brew, possibly because of the invert sugar and caramel added to the all-malt mash or the German hops used for dry hopping. Whatever the reason, it is an instantly recognizable bitter brew, not by any means to everyone's taste but available in traditional form in two-thirds of the pubs, which are widely dispersed from Wakefield to the north Humberside coast and from Retford to Selby. There is also some free trade (there was very little before the take-over), and Darley's beers can be found in Ward's houses and also in Vaux houses as far away as Newcastle. Since Darley's have room to expand, there is every reason to believe that, even in these days of rationalization, Vaux will continue to brew in Thorne – though that thought must be worrying for supporters of Ward's, the fellow Vaux subsidiary, which has a rather more confined site only twenty-five miles away in Sheffield.

SAMUEL SMITH'S
Samuel Smith's Old Brewery (Tadcaster) Ltd, The Old Brewery, Tadcaster, North Yorkshire.
Pubs: 300. Very substantial and widespread free trade. Beers: 4X Mild (1033), Tadcaster Bitter (1035), Old Brewery Bitter (1039), Museum Ale (1047).

Yorkshire's oldest brewery, with its long-standing and deserved reputation for 'beer from the wood', Samuel Smith's brew what is still one of the best-known real ales, Old Brewery Bitter. Yet this beer dates only from 1974, when no fewer than four traditional draught beers were withdrawn, to be replaced by only two, and for seven years after 1977, when 4X best mild was withdrawn in draught form, it was Sam Smith's only traditional ale.

The brewery was founded in 1758 and in 1847 was bought by Samuel Smith; by 1879 it was in the hands of his grandsons, Samuel and William. When Samuel II died, an argument developed between his son (also Samuel) and William, and this was only resolved when William pulled out of the firm altogether, building another brewery a mere 200 yards away: this survives, too, as the much larger John Smith's Brewery, part of the Courage group. Sam III, at the tender age of twenty-one was left to run the Old Brewery – with such success that a century later, still run by the Smiths (Humphrey and Oliver), fourth generation descendants of Sam II, the firm has around 300 pubs and a vast presence in the free trade.

The brewery is run on intensely conservative lines, with the consequence of an image of traditionalism which is all-pervasive. The brewing-water is still drawn from the original well at the brewery, sunk more than two centuries ago. Some of the coppers are over a hundred years old and are amongst the oldest still in use in Britain, the strain of yeast dates from the closing years of the nineteenth century, and the beer into which it is pitched is fermented in slate Yorkshire squares. Each of these fermenting vessels, described as 'roofed' because there is a partition at about two-thirds height, can hold seventy-five barrels (21,600 pints) of beer, and when the green beer is first pumped into the vessel it froths through an opening in the roof and then subsides as the fermentation settles down. The result of this unusually vigorous fermentation is a beer with a unique and distinctive palate.

The accent on tradition is just as strong after the fermentation ceases and the beer is racked into casks, for all of Samuel Smith's traditional draught beer is stored in wooden casks, and they are one of the very few remaining breweries to employ their own cooper. Indeed, the launch of a new beer (Tadcaster Bitter) in 1984 was delayed because of the problems encountered in acquiring sufficient new oak casks, and eventually special cask-making machinery was bought from a defunct cooperage in Sheffield to build up a stock of wooden casks, regarded as a sound investment not only because of their almost unlimited lifespan but because of the flavour they impart and the traditional image they convey.

Smith's keep their own dray, too, and they also took on the responsibility, when they agreed to supply Melbourn's pubs, for the brewing museum at Melbourn's former brewery in Stamford. Despite the emphasis on tradition, though, Samuel Smith's is run not by romantics but by hard-nosed businessmen who recognize the growth in lager (hence their lager plant, next to the Old Brewery, devoted to brewing Alpine lager under licence from Ayingerbrau) and the need for brewery-conditioned beers (hence their considerable presence in the northern club trade) and who have also decided to convert the entire tied estate to managed rather than tenanted houses, against the trend in the industry.

The range of traditional beers has recently increased rapidly, but the undisputed leader is still Old Brewery Bitter (1039), a very full and malty beer which can have a rather cloying taste. It is certainly highly individual in character, despite a reduction in gravity from 1040.9 to 1038.9 degrees in 1979, just five years after it was introduced. It is based on a recipe from the 1920s which was rediscovered in the brewery. Tadcaster Bitter (1035), introduced in September 1984, is completely different – very light-coloured, delicately flavoured and with a much more hoppy taste. Museum Ale (1047), named after the Stamford brewery museum and initially made available mostly in London and the East Midlands, is a premium brew introduced in 1985, together with 4X Mild (1033), a light and pleasant mild.

In addition to an almost nationwide free trade there is a tied estate of about 300 houses, still growing as Sam Smith's continue to purchase pubs. In addition to the contract to supply Melbourn's pubs, Smith's bought some of Ruddle's tied estate when that was sold off in 1978, and since April 1980 they have increased their stock of London pubs from none to nineteen, including the Swiss Cottage, reportedly bought for a record £1.7 million, and the Gazebo in Kingston, a brand-new pub opened in 1982.

TAYLOR'S
Timothy Taylor & Co Ltd, Knowle Spring Brewery, Keighley, West Yorkshire.

Pubs: 29. Beers: Golden Best or Bitter Ale (1033), Mild (1033), Best Bitter (1037), Landlord (1042), Porter (1043), Ram Tam (1043).

Taylor's is a typically Tyke business, staunchly independent and very sensitive, indeed prickly, in response to enquiries about its wellbeing. It is also notably proud of the string of awards gained by its 'championship beers' – the firm's notepaper is weighted down by the list of achievements. Yet, until a recent widening of Taylor's free-trade activities, it was necessary to travel to Keighley or its environs to seek out the much-talked-about but infuriatingly elusive beers.

Taylor's have been brewing at the Knowle Spring Brewery (itself a fairly elusive headquarters, unsignposted to deter would-be visitors) since 1863, although the present chairman's grandfather, Timothy Taylor, started brewing five years earlier, at a small brewery in Cook Lane, Keighley. The tied estate has been built up gradually, with pubs bought from Truman's, when the London brewery closed their Burton offshoot, and from local brewers Aaron King (who gave up brewing in the Old Brewery, Cook Lane, in 1946) and Parker's of Howarth, whose Clarendon Brewery ceased production when the firm went into voluntary liquidation in 1966. One pub, the Red Lion in Colne, is leased from Sam Smith's in return for Taylor's pubs stocking Smith's lager; two new pubs were built in the 1970s – the Knowle Arms and the Timothy Taylor, both in Keighley, where a perhaps excessive fifteen of the twenty-nine pubs are located; the latest addition to the tied estate saw Taylor's move into York in 1984, taking over the Brown Cow in Hope Street, again from Sam Smith's.

The brewery is totally traditional, with an open copper and stainless steel fermenting vessels, and the beers are produced from an all-malt mash, with no adjuncts, and whole hops. The brewing liquor is pure and very soft well water, pumped from a vast natural subterranean lake below the brewery, reputedly filled by underground streams emanating from Malham Tarn. The result is a surprisingly large range of draught beers, including a recently introduced porter.

By far the best known of Taylor's beers, though certainly not its biggest seller, is Landlord (1042), a popular, malty and quite full-bodied bitter. Both Ram Tam (1043), a dark and quite sweet winter brew, and Porter (1043), have noticeably similar gravities to Landlord, although the porter is said to be based on a nineteenth-century recipe found in old brewing books. The best-selling beer is Best Bitter (1037), a pleasant though not especially distinctive drink, although until the 1970s the most popular order in Taylor's pubs was Golden Best (1033). This light, rather bland beer is also known as Bitter Ale, Golden Mild and Light Mild, and a darker version of it is racked as Dark Mild (also 1033).

The future of the brewery, given the popularity of the beer in Taylor's twenty-nine pubs and an appreciative and far-flung free

A Taylor's beer label

trade (including pubs in Edinburgh and Herefordshire), seems assured, particularly whilst the present chairman, Lord Ingrow, is in control. Previously, take-over offers for the company simply elicited no reply. The shares are still entirely family-owned, and although Lord Ingrow has no immediate heir, his philosophy remains positive: 'We might have to reorganize a little, but we will survive.'

THEAKSTON'S
T. & R. Theakston Ltd, Wellgarth, Masham, Ripon, North Yorkshire. Also brew at Bridge Street, Carlisle, Cumbria.
A subsidiary of *Matthew Brown*, with twenty pubs and an exceptionally extensive free trade. Beers: Best Bitter (1037), XB (1045), Old Peculier (1058).

Theakston's was one of the shooting stars of the real-ale boom of the 1970s, but it has taken the company until 1984 to overcome its resulting financial problems, and to do so it has had to forfeit its independence, bringing fears for the future of the breweries, especially Carlisle, although Matthew Brown have so far been very positive in extending the Theakston tied estate and promoting the beers in their own pubs.

In 1972 Theakston's supplied beer to fewer than fifty outlets, none of them more than seventy-five miles from their Masham brewery (built in 1875, though the firm was founded by Robert Theakston in 1827 and was involved in only one take-over, that of their Masham neighbours Lightfoot's, whose Wellgarth Brewery was closed in 1919). But Theakston's had an ideal product for the burgeoning real-ale market in Old Peculier, a strong dark ale with a unique flavour which spearheaded the company's rise to prominence. Over a thousand free-trade outlets throughout Britain were selling Theakston's beers by 1976, and such was the company's confidence that it bought the former State Management Scheme's Carlisle brewery (with a capacity of up to 2,000 barrels a week) to supplement the Masham brewery, whose capacity was just 250 barrels a week.

Despite the pace of expansion, however, the Carlisle brewery was brewing at only twenty-five per cent capacity in 1976, bringing cash-flow problems which were alleviated only by selling seven of the fourteen tied houses and allocating thirty-five per cent of the company's shares to a London management firm, Farvise, which provided much-needed capital. Three more pubs

A distinctive Theakston's beer mat

were put up for sale in 1977, confirmation that the Yorkshire firm had overstretched itself in acquiring the Carlisle brewery. But within four years Theakston's were buying pubs again, though this was with the backing of the London Trust, an investment company which took a forty-eight per cent holding in 1978, largely to finance the building-up of production at Carlisle.

More changes took place in 1983, when Michael Abrahams, a Yorkshire textiles millionaire, offered to buy twenty-nine per cent of the shares at 40p a share. Some of the board members turned instead to Matthew Brown, who agreed to make a 64p-a-share offer, while others objected to the prospect of take-over by a brewing company and persuaded the whisky distillers William Grant to make an 88p offer (worth £3.7 million). Paul Theakston, speaking for his eleven per cent shareholding and the forty-eight per cent held by London Trust, which he had agreed to buy, rejected the higher offer (and even higher bids rumoured to have been made by Samuel Smith's and others) and eventually prevailed over the other directors, with Matthew Brown gaining control in July 1984.

Theakston's breweries and beers have been subjected to less dramatic changes, although Old Peculier has been slightly weakened, two milds have been withdrawn and a very good new premium bitter, XB, has been introduced. Old Peculier is named after the eighteenth-century Peculier of Masham, a parish free from diocesan jurisdiction, and it has enjoyed a remarkable recovery from comparative obscurity to international fame. As recently as 1969 it was brewed only once every three months, in batches of thirty-three barrels; ten years later nine barrels a week were brewed, and since then the recession has marginally reduced that figure. Most of the 'OP', brewed from malt, sugar, malt extract, caramel and flaked maize, was fermented until recently in slate squares (now more stainless steel fermenters have been added) and it is dry-hopped to add a touch of bitterness to the otherwise rather cloying taste of the 1058 brew (1060 in the 1970s). XB (1045), introduced in 1982, is an extremely good premium bitter, full of flavour, which now accounts for over half the 200 barrels a week brewed at Masham.

The sole traditional beer from Carlisle is Best Bitter (1037), very pale and distinctive and accounting for the bulk of the very respectable 1,500 barrels a week brewed in the historic Caldewgate Brewery, much of which is at least 200 years old and which is little changed since the fermenting rooms were updated in the 1960s. Best bitter has both its admirers and its detractors, but it is now a consistent, easily recognized and widely available pint – the more so since the Matthew Brown take-over; and provided the Carlisle brewery survives, it seems likely to grace the bars of many free houses for some time to come.

VAUX
Vaux Group plc, The Brewery, Sunderland, Tyne & Wear.
Pubs: 450. Beers: Sunderland Draught Bitter (1040), Samson (1042).
Also owns *Lorimer & Clark* (Edinburgh), *Darley* (Thorne) and *Ward's* (Sheffield).

Vaux, with four breweries in Britain, a major chain of hotels, a wine and spirits merchant, an off-licence chain and a share in Tyne-Tees television, is entitled to be regarded as a major regional company. Indeed, the Swallow Hotels, of which there are thirty-one and which now earn a significant proportion of the group's profits, give credence to Vaux's claim to a national presence – a claim which is underlined by the recent purchase of five pubs in London.

The core of Vaux's business is, however, in north-east England, hard hit by recession but still big beer-drinking territory. Sunderland thirsts have been quenched by Vaux since 1806, when Cuthbert Vaux became a partner in a brewery in Moor Street. Thirty-one years later two of his sons joined him in a new brewery venture, first on Matlock Street and then on Union Street. In 1875 C. Vaux & Sons had to move yet again, to the present site, this time because the Union Street brewery was required as the site of Sunderland's railway station. The new brewery was soon known throughout northern Britain for its 'nourishing dry stout' – so much so that bottling plants were established in Glasgow and Leeds – but disaster befell this brew during the First World War, when sugar-rationing forced a break in production. By the time restrictions were lifted, a certain Irish stout had stolen the market.

Recession in the 1920s saw the amalgamation of a series of north-eastern breweries, and Vaux & Sons decided to merge with North East Breweries (also of Sunderland) in 1926 to create Associated Breweries. The new firm expanded further afield in the 1930s, notably with the purchase of the Berwick Brewery in 1937, and during and after the Second World War it ventured further south (Hepworth's of Ripon in 1947), crossed the Pennines (Whitwell, Mark & Company of Kendal and its thirty pubs, also in 1947; the brewery finally closed in 1968) and even penetrated into Scotland (the Coldstream Brewery, 1943, and Lorimer & Clark of Edinburgh, 1946). The Scottish presence was consolidated with the acquisition of the Edinburgh firms of Thomas Usher (170 pubs) and Steel Coulson (sixteen pubs) in 1959, and even a brewery in Perth two years later. Finally the 1970s saw Ward's of Sheffield and Darley's of Thorne join the group – both are still brewing – and in a surprising deal in 1980 all the Scottish pubs were sold to Allied Breweries, leaving Vaux's surviving Scottish brewery, Lorimer & Clark to supply its excellent 'Scotch' ale to Vaux pubs, and its other beers to the Scottish free trade.

Although, as a major public company, Vaux is now largely owned by financial institutions (of which Britannic Assurance, with six per cent, is the largest), there is still a strong continuity of family interest in the company. The present chairman, Paul Nicholson, is a great-grandson of Cuthbert Vaux, and in addition Peter Vaux is a non-executive director. The company has faith in certain traditions – most obviously the use of horse-drawn drays to make local deliveries. There are about twenty heavy horses, Percherons and Dutch Gelderlanders at present, and they deliver some 200 tons of beer a week to tied houses in Sunderland.

Not a great deal of the beer is served traditionally in the pubs, for only some 170 of the 450 Vaux pubs sell real ale. The two cask-conditioned brews from Sunderland are the pale, rather uninspiring Sunderland Draught Bitter (1040) and the much fuller-bodied, darker and tasty Samson (1042), a really robust and distinctive premium bitter. Lorimer's Scotch (the 70 shilling ale from the Caledonian brewery in Edinburgh) is very popular in Vaux houses, and the brewery prides itself on being the only English brewer to offer a genuine 'Scotch' ale brewed in Scotland.

WARD'S
S. H. Ward & Co Ltd, Sheaf Brewery, Eccleshall Road, Sheffield.
A subsidiary of *Vaux*, running 102 pubs. Beer: Sheffield Best Bitter (1038). Also sells Dark Mild from Darley's, another Vaux subsidiary.

Having sailed close to the wind twice in the 1880s, first failing to settle with their creditors and then finding the banks foreclosing on them after a disastrous speculation in the shipping business, Ward's survived another ninety years before their independence finally ended, with their absorption into the Vaux group in 1972–3. Ward's clearly believed their days were numbered after the merger mania of the 1960s, and the then chairman, Wilfred Wright, reputedly approached Vaux, the company with which he considered the effects of a take-over would be least damaging – for the company, its staff and its customers. Since then there have been two major changes, with the end of the blending and bottling of wines and spirits and the closure of the beer-bottling department, but at least real beer survives – though there is now only one traditional brew, and its future could be threatened if Vaux choose to concentrate production at their other South Yorkshire subsidiary, Darley's.

The brewery, originally the Sheaf Island Brewery in Effingham Road, dates from about 1840, and some forty years later was run by a Mr Kirby, who was unable to settle his bill for barley with George Wright, great-grandfather of the present managing director. Wright stayed on at the brewery and was taken into partnership, but Kirby, Wright & Company soon ran into trouble when George Wright acted as guarantor for a shipping company which lost a clipper at sea. The banks foreclosed immediately, and the company would have collapsed but for the intervention of Septimus Henry Ward, who secured its financial future. During a period of expansion the Albion Brewery and the adjacent Soho Brewery were acquired, and three breweries were run until the Soho Brewery was completely modernized (and renamed) and production concentrated at one site.

Ward's originally sold their beer solely to free houses but have gradually acquired an estate of around a hundred pubs, mainly in or close to Sheffield. One interesting exception is the Puzzle Hall in Hollins Mill Lane, Sowerby Bridge, near Halifax. Dating back to the early seventeenth century, when it was built as a private dwelling, the Puzzle Hall was a home-brew pub by 1850, and in 1905 a tower brewery was built; beer labels for Platt's Sowerby Ales still survive. Ward's acquired the property in 1935.

Puzzle Hall ales are no more, and the same can be said of a number of Ward's own beers. Quite apart from the bottled beers,

A label for Home Brewed Stout from the Puzzle Hall Brewery, taken over by Ward's in 1935

three draught brews have been discontinued recently. Ordinary bitter was the first to go, at the end of 1976. Sheaf Ale, a light low-gravity bitter introduced, partly as a replacement, in 1980, lasted just three years before declining sales led to its withdrawal. Ward's dark and malty mild had a longer run but it, too, bit the dust in January 1984, replaced at first by a stronger porter-style dark ale from Darley's (promoted as 'more than mild, better than bitter'). This new beer was itself replaced by dark mild, also from Darley's, later in 1984. The only traditional brew from Ward's is now Sheffield Best Bitter (1038), a popular, full-flavoured and slightly sweet beer sold in traditional form in about two-thirds of the pubs.

11 Scotland

Scotland's once-flourishing brewing industry, centred on famous brewing towns and cities such as Alloa and Edinburgh (which alone had eighteen breweries in 1960), was devastated in the 1960s and there are now just eight breweries in the whole country. Five are controlled by the Big Six: Bass have two, Tennent's in Edinburgh and Glasgow; Allied controls the Alloa Brewery; Watney's has Drybrough's in Edinburgh, and Scottish & Newcastle still brews at the Fountain Brewery in the Scottish capital, having closed its Holyrood Brewery in 1985. One regional brewer, Vaux, has a Scottish outpost, Lorimer & Clark, in Edinburgh. And there are just two 'independent' breweries, Maclay's (in which Bass has a major share stake) and Belhaven, a family brewery until 1972 but with a more colourful recent history.

Disaster befell Scotland in 1960–61, largely as a result of the ambition of the Canadian entrepreneur Eddie Taylor and his Northern Breweries Ltd. Formed in 1959 as the vehicle for the merger between Hope & Anchor of Sheffield, Hammond's of Bradford and John Jeffrey & Co of Edinburgh, the company turned its attentions on the Scottish independent brewers in the next two years, before, as the renamed United Breweries Limited, becoming embroiled in the Bass Charrington empire via a merger with Charrington. Amongst its Scots victims were the Edinburgh breweries of Aitchison's and Murray's, Aitken's of Falkirk, the two Alloa firms of James Calder and George Younger, and John Fowler of Prestonpans. All these brewers' pubs eventually passed to Tennent Caledonian Breweries as Charrington rationalized its Scottish interests in the 1960s.

Scottish & Newcastle Breweries, its Scottish base born out of a merger between the Edinburgh firms of William Younger (founded in 1749 at Leith and running the Abbey and Holyrood Breweries) and William McEwan (of the Fountain Brewery) in 1931, had also steadily bought out competitors, and the other major brewers were not to be outdone. Watney's collected Gordon & Blair in 1962 and Drybrough's in 1966; Whitbread bought Archibald Campbell, Hope & King in 1967 and, amid storms of protest, closed the brewery three years later. Allied Breweries owe their Scottish presence largely to Samuel Allsopp's take-over of Archibald Arrol's Alloa brewery in 1930.

Four of the breweries controlled by the Big Six produce traditional beer, though most of it is unexceptional. The other surviving brewers do rather better, with Lorimer & Clark producing some really excellent beer, and both Maclay's and Belhaven brewing a fine 80/- Export ale. Hardly any dark mild beer (referred to as 60/- Light north of the border) is brewed, and the preference is for slightly sweet and malty bitter beers. The beers are described, as with the other Scottish breweries, in terms of their historical liability for excise duty. The best-known beer from the new breweries, Broughton Brewery's Greenmantle Ale, follows this style. One other brewery should be noted: at Traquair House, near Peebles, the laird rediscovered an eighteenth-century brewhouse in 1965 and has since been brewing a very strong (and usually bottled) Traquair House Ale, and more recently a strong draught beer, Bear Ale.

BELHAVEN
Belhaven Brewery Group plc, The Brewery, Dunbar, East Lothian, Scotland.
Pubs: 25. Very extensive free trade. Beers: 60/- Light (1031), 70/- Heavy (1036), 80/- Export (1042), 90/- Strong Ale (1070).

After a long and uneventful history, tiny but still independent until 1972, Belhaven entered a decade or more of rapid change, complete with financial crises and boardroom upheavals. Thankfully the very individual beers have escaped the turmoil pretty well unscathed, and they are still widely available in Scottish bars.

The brewery was founded by 1719 and was under the control of the Hunter family for over a century until 1972, when death duties forced Sandy Hunter to sell the company – then known as Dudgeon & Co – to Clydesdale Commonwealth Hotels for just £82,000 (£12,000 for the brewery and £10,000 each for the seven tied houses). CCH, formed in 1968, had already gained a reputation for quickfire buying and selling of assets, but it held on to the Belhaven brewery, which brewed fewer than 4,000 barrels in 1972 but within two years had increased output more than fourfold and had reached 30,000 barrels by 1978. The brewery quickly dominated the rest of CCH, which renamed itself the Belhaven

A Belhaven beer mat

Brewery Group and sold off most of its British hotels and its travel company, leaving just Belhaven (now with twenty-five tied houses) and a hotel in Bermuda. Other deals included a £250,000 loan from Allied Breweries and the purchase of Ashpoint, a packaging firm.

These developments were quickly overshadowed by the events of 1979, when the company had five chairmen in just six months as a result of boardroom squabbles. By the end of the year Eric Morley, formerly of Mecca, was co-chairman, and the activities of the brewery group included running the Miss World competition. There were still packaging and construction interests, too, but the loss-making hotel in Bermuda was sold. Yet financial problems continued to beset the group – profits of £600,000 in 1981 slumped to just £4,000 in the following year – and a holiday camp in East Anglia, together with a Spanish hotel, was next to be sold. In 1983 Miss World departed and Ashpoint was also disposed of, and in 1984 Nazmu Virani bought a twenty-seven per cent stake and became chairman and chief executive.

By this time Belhaven had no pubs left, but Virani quickly began a pub-buying programme and also forged a link with Courage, distributing the Big Six brewers' beer in Scotland. He also began looking for a brewery in southern Britain, acquiring a small stake in Llanelli brewers Buckley's, considering the purchase of Bateman's, buying a London pub and chain of off-licences and expanding free trade. All these activities, significantly, were connected with brewing, in which business Belhaven had continued to trade profitably throughout their times of trouble.

Considerable alterations have been made to the brewery in recent years in order to meet demand, though parts still have the two-foot-thick stone walls believed to have been built by the monks in the fourteenth century. Capacity has been increased with an enlarged mash tun and new coppers (a ninety-six-year-old copper was replaced in 1984) and more fermenting vessels, and the increasing need for space has seen the end of malting at Belhaven (discontinued in the 1970s) and the purchase of adjacent properties in Dunbar.

The four beers, normally dispensed by air pressure in Scotland, retain their very individual character, which is attributed mainly to the Dunbar well water and the use of high-quality raw materials. 60/- Light (1031), which is dark and malty, sells fewer than twenty barrels a week, yet the brewery are prepared to vary its sweetness and colour for individual outlets. 70/- Heavy (1036) sells little more than the Light, which is surprising since it is a well-hopped and very palatable brew. But 80/- Export (1042) is several times more popular; rich, heavy and quite dark, it is a very smooth, full-bodied ale. The fourth traditional beer is 90/- Strong Ale (1070), very rich and sweet, which is regularly brewed but served in only a few outlets.

LORIMER'S

Lorimer & Clark Ltd, Caledonian Brewery, Slateford Road, Edinburgh.

A subsidiary of *Vaux*, now owning no tied houses and selling to the Scottish free trade, in addition to supplying 70/- to Vaux houses in north-east England. Beers: 60/- (1030), 70/- (1036), 80/- (1043), Caledonian Strong Ale (1077).

Lorimer's is a precious survival in the Scottish brewing scene, and an extraordinary one at that. Having lost its independence as long ago as 1946, to Sunderland-based Vaux, the company lost its 214 tied houses in 1980 when Vaux sold them to Allied Breweries. Fortunately Vaux chose to shut down their other Edinburgh brewery, Thomas Usher, producers of much-inferior ale, and Lorimer's have journeyed on, introducing new and distinctive ales to an enthusiastic Scottish free trade. Their true salvation, however, lies in the praiseworthy insistence of Vaux that the beer they sell in their own pubs as 'best Scotch' should actually be brewed in Scotland (attempts to brew it in Sunderland having proved unsuccessful).

The brewery dates back to 1869, when George Lorimer, who had begun his career in the brewing business only four years earlier, at the age of nineteen, formed the company with Robert Clark, previously a brewer at the large Edinburgh firm of Alexander Melvin. Lorimer's expanded steadily, searching for outlets in both western Scotland and north-eastern England, and it was this latter circumstance which led to a trading agreement with Vaux & Associated Breweries in 1919, and take-over in 1946. Vaux continued to expand their activities in Edinburgh, buying both Thomas Usher (whose 170 pubs were supplied from their Park Brewery) and the much smaller Steel, Coulson's Croft-An-Righ Brewery in 1959. From 1960 Lorimer's beer was unobtainable in Scotland; only Usher's ales, widely regarded as Scotland's worst, were supplied. Lorimer's were therefore in the odd position of having a fine traditional brewery – its most noteworthy feature the unique survival of three coal-fired open coppers (one of them still the original, the two others replaced in the 1980s) – but no local trade.

Now Lorimer's is well and truly back as one of Scotland's finest brewers. The very pleasant, malty 70/- ale (1036), sold in England as Lorimer's Best Scotch, still accounts for most of production but has been joined by three others. The closure of Usher's brewery coincided with the launch of one of these, 80/- (1043), which is sold only in the Scottish free trade. It has quickly and deservedly gained an excellent reputation as a balanced, flavoursome premium bitter. And in 1984 the smooth, dark 60/- (1030) was added, though inevitably this welcome introduction of a low-gravity dark ale has had a limited impact, with the beer to be seen in only a few bars. Finally, Caledonian Strong Ale (1077), which was previously all sent in bulk to Sunderland for bottling by Vaux, is now occasionally available on draught.

Although the brewery remains dependent on Vaux's will to continue brewing Lorimer's Best Scotch in Scotland, there are now a healthy number of free-trade accounts, mainly in Edinburgh and central Scotland but with outposts in Aberdeen and the Scottish Borders. Many fine Edinburgh drinking-shops are now further enhanced by the fact that they stock Lorimer's ales – notably Leslie's Bar in Ratcliffe Terrace, a superb panelled Victorian bar with a carved snob screen and ornate ceiling.

MACLAY'S
Maclay & Co Ltd, Thistle Brewery, Alloa, Scotland.
Pubs: 28. Considerable free trade. Beers: 60/- Light (1030), 70/- Heavy (1035), 80/- Export (1040).

Given the devastation of the Scottish brewing industry in the 1960s and the fact that one of the Big Six has for some time had a substantial share stake, the survival of Maclay's as an independent force in Scottish brewing is little short of miraculous. Yet Maclay's are still brewing in Alloa, once famous for its breweries. There were seven in the town at the turn of the century, including Archibald Arrol's (sucked into the Allied Breweries combine via take-over by Samuel Allsopp of Burton-on-Trent) and George Younger, of the Meadow and Candlerigg Breweries, acquired by Northern Breweries in 1960 – but now only two remain. Maclay's only local rivals are now Allied, who have kept open the Arrol's plant almost exclusively to brew lager.

James Maclay entered the brewery trade in this small Clackmannanshire town in 1830, when he leased the Mills Brewery; previously, though a native of Alloa, he had been a clerk at the Devon Ironworks in nearby Tillicoultry and then an accountant at the long-vanished Hutton Park Brewery in Alloa. He was the lessee of the Mills Brewery for thirty-nine years, leaving only when business had expanded to the point where it was necessary to build a brand-new brewery, the Thistle Brewery at the east end of the East Vennel. The effort involved in commissioning the new brewery left its mark, however, and James Maclay died in 1875. He was succeeded by his two sons for a time, but the Maclay family connection was severed in 1896 when Alexander Fraser bought the firm. Fraser is credited with further expanding trade, so much so that the premises soon occupied almost the whole of one side of Old High Street.

Maclay's have fallen under the influence (reportedly not very direct) of Bass by an unusual route. The company's operations were once much more extensive, with major exports to Ireland, Malta and India catered for by a bottling hall in London, and an estate of around sixteen pubs around Durham and Newcastle. Transport difficulties after the Second World War precipitated the sale of these pubs to Hammond's of Bradford, and when Hammond's – who had taken a stake in Maclay's at the time of the pub sales – were successively absorbed into Northern Breweries, Charrington and then Bass Charrington, the Maclay's holding was retained and, indeed, increased. Now Bass have thirty per cent of the shares and a seat on the board, but Maclay's appear

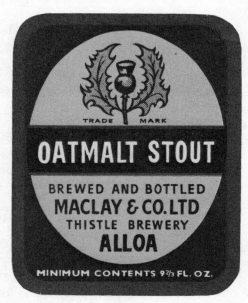

A label for Mackay's well known 'Oatmalt Stout'

determined to remain independent. The ultimate holding company is Fraser & Carmichael (Holdings), the name a throwback to earlier days, and the major shareholders are reputedly opposed to any loss of independence.

Alloa's ales acquired quite a reputation over the years, and they were described as 'of great celebrity not only in Scotland but in distant lands' in the 1860s. The present contribution made by Maclay's comes in three forms. 60/- Light (1030) is a dark beer, the name referring to its light, refreshing flavour. 70/- Heavy (1035), in contrast, is a consistent and fairly hoppy brew which has traditionally sold best in the coalmining areas of Fife but which is now fairly widely available in both tied and free trade; and 80/- Export (1040) is a quite strong, well-balanced bitter. Sadly, Old Alloa Ale, which was an excellent rich winter ale brewed to a gravity of around 1065, is no longer sold.

The tied houses are very widely distributed throughout central Scotland, with only one in Alloa and others dispersed as far as Perth (the Hal o' the Wynd), Edinburgh (the excellent Southsider) and Saltcoats on the west coast. Traditional ale is available in two-thirds of them, almost invariably served by air pressure via the equally traditional Scottish tall font – Maclay's having no doubt that this is the optimum method of serving their beers. But, this being Scotland, tied houses account for only a small proportion of Maclay's trade, and the beers are widely available in Scottish free houses.

12   Offshore Breweries

Britain's island breweries, except those on the Isle of Wight, have generally been well insulated against take-over and closure. In the Channel Islands the four established brewers, two each on Guernsey and Jersey, have generally been more at risk from German invaders than merger maniacs. The Isle of Man has seen the five breweries which existed at the beginning of the twentieth century reduced to two, with Castletown absorbing Clinch's, whose brewery still stands on North Quay, Douglas, and Okell's now supplying the Heron & Brearley pubs. The only external predator was Ind Coope, who bought Kewley's brewery in Douglas as long ago as 1902 (although Boddingtons' were also briefly involved in the island trade, as explained on page 175). Bass, though, have a thirty per cent shareholding in Castletown Brewery, a legacy of the deal which saw Hope & Anchor Breweries of Sheffield fleetingly in control of the Manx brewery in the 1940s.

The Isle of Wight has suffered rather more. Eleven breweries saw in the twentieth century, but within twenty years six had ceased trading, and one, Sweetman's of Ryde, had been taken over by the island's major brewers, Mew, Langton & Company, whose Royal Brewery in Newport had been founded in 1814. Sprake's Chale Brewery was closed in 1934, six years after it had been bought by the Portsmouth brewers Brickwood's, and in 1953 the Shanklin Brewery brewed its last pint; finally Mew, Langton was bought by Strong's of Romsey, and after Strong's had themselves surrendered their independence, to Whitbread, the Newport brewery was closed in 1969, leaving Burt's as the island's only brewers.

The majority of the offshore breweries produce traditional beer, and some of it is excellent – Castletown Brewery's bitter, Burt's VPA and the Guernsey Randall's Best Bitter are good examples. These three breweries, together with Okell's and Guernsey Brewery, make real draught beer widely available. The blackspot is Jersey, where Ann Street have shown no interest in anything but keg beer, and Randall's, after briefly re-introducing it, reverted to keg because of 'lack of demand'.

Guernsey

GUERNSEY BREWERY
The Guernsey Brewery Co (1920) Ltd, South Esplanade, St Peter Port, Guernsey.
Pubs: 35. Free trade throughout the island. Beers: LBA Mild (1037), Draught Bitter (1045).

One of the most delightfully situated breweries in the country, right on the water's edge in St Peter Port and with marvellous views of Castle Cornet and the harbour, Guernsey Brewery, the larger of the island's two breweries, produces excellent and quite strong draught beers, though most of its sales derive from keg and lager. Tradition, however, does still play a part: the brewery still retains the horse-drawn dray which it bought in 1912, and uses it for deliveries in St Peter Port during the summer.

Brewing at South Esplanade began in 1856, when John le Patourel built the curiously named London Brewery there, alongside his maltings and kilns for roasting chicory. Thirteen years later le Patourel let the brewery to Messrs Richings & Baxter as tenants, and later Richings traded alone, advertising his prize medal ales as 'warranted free from adulteration and superior to any Ales imported' in the 1870s. In 1886, however, the business reverted to le Patourel's son-in-law, Thomas Girling, and three years later H. J. Flambe & Co became tenants, brewing there for five years and then going bankrupt.

A measure of stability was achieved in 1895, when Messrs Schreiber & Skurray leased the property, forming the Guernsey Brewery Co with, oddly enough, registered offices at Morland & Co in Abingdon, home town of Skurray. The present title was assumed in 1920, when the company was re-registered in Guernsey for tax reasons, and at the same time le Patourel interest was at last bought out. Three years later the present, much larger brewery was opened.

The war years of 1940–45 were desperately difficult, with the sources of supply for raw materials cut off. Despite the Occupation the brewer, Roy Higgs, displayed extraordinary inventive-

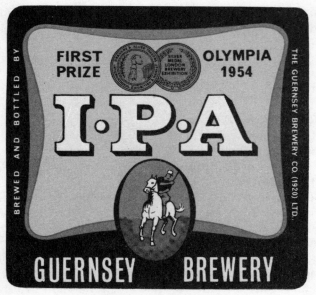

A beer label from the Guernsey Brewery

ness, first reducing the original gravities to conserve supplies of malt, then producing a process beer from invert sugar, using sugarbeet and even parsnips and finally making a non-alcoholic 'hop ale' from hops, saccharine and grain spirit. Even after Higgs was deported to Germany, and parts of the brewery were commandeered as a hospital, production of the hop ale continued until the end of the war. Deliveries were made by horse and cart, by a truck pulled by an ox, and even by hand-drawn trucks.

Between 1947 and 1954 the brewing plant was completely replaced, a new bottling line was added and the tied estate was substantially increased. A new keg store was built in 1972, in recognition of the growing demand on the island for keg beer (which now represents seventy per cent of Guernsey Brewery's production), and mainland lager imported in tankers began to be kegged at the brewery. Subsequently Stein lager brewed under licence from Thwaites of Blackburn has been produced, in Britain's smallest conical fermenter, standing in the brewery yard.

In 1977 Captain Schreiber's son-in-law, Captain Richard Symons, who had been managing director since 1948, decided to retire, and in the absence of interest from within the family in carrying on the business it was offered for sale and bought by Bucktrout Limited, a Guernsey company which still owns the brewery.

Guernsey Brewery produce about 10,000 barrels of beer a year, including two traditional draught beers, although these account for only a minority of output. The beers are relatively strong because of the Channel Islands' unique method of levying duty (as described on p. 232). LBA (1037) is a fine, rather heavy medium-dark mild which has quite a sweet taste even though it is generously hopped. The initials stand for London Brewery Ale, a reference to the original name of the brewery. LBA, very popular with the islanders, accounts for almost half of draught beer production. The rest is Draught Bitter (1045), a very full-flavoured and hoppy strong bitter which is in about half of the tied houses, having been reintroduced in the late 1970s (LBA has always been available in real draught form). Keg beers are filtered versions of the draught brews, and Pony Ale and IPA are the same beers in bottled form. Two bottled-only brews are Stout and Brown Ale.

RANDALL'S (GUERNSEY)

R. W. Randall Ltd, Vauxlaurens Brewery, St Julians Avenue, St Peter Port, Guernsey.
Pubs: 17. Island-wide free trade. Beers: XX Mild (1037), Best Bitter (1044).

The Vauxlaurens Brewery, separated from St Peter Port's colourful harbour by the steep, tree-lined St Julians Avenue, is thought to date back to the seventeenth century, though the earliest known owner is Joseph Gullick, who owned not only the brewery but also a brickworks and a rope walk in the 1860s. Gullick sold the brewery in 1868 to R. H. Randall, recently arrived from Jersey (and from the same family as the Jersey brewers).

The Randalls have controlled the brewery since then, and at one time also ran a mineral-water factory, although this was sold in 1921 to the Guernsey Aerated Water & Jam Company. R. H. Randall died in 1899 and was succeeded by his son R. W. Randall, whose initials survive today; R.W.'s sons were taken into the business in 1927 and 1930, and in 1929 the present private company – all of whose shares remain in the hands of direct

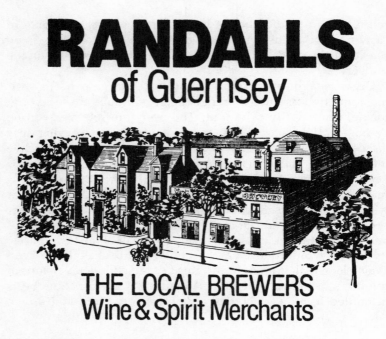

A beer mat advertising the wares of Randall's of Guernsey

descendants of the original R. H. Randall – was formed. In 1954 the fourth generation began to join the company, now run by Paddy Randall.

The beer is produced in a traditional tower brewery, though much of the structure is comparatively recent, including the brewhouse and bottling store, rebuilt in 1950–52. There are two draught beers, including an excellent strong mild – its high strength compared with ordinary British milds being a consequence of the Channel Islands' excise system, in which duty is levied on quantity produced rather than the strength of the beer, so that there is no tax advantage to the brewer in keeping the original gravity low.

XX Mild (1037) is a very full-bodied, dark and rather sweet drink available in traditional draught form, dispensed either by handpump or direct from the cask, in a dozen or so outlets, including some free-trade pubs. Best Bitter (1044), which for some time was sold traditionally in only one outlet – although it was never actually discontinued – is now also in a handful of Guernsey bars; it is a well-hopped, refreshing beer which deserves to be more widely known. Altogether cask-conditioned beer accounts for only three or four per cent of Randall's production, but its future seems reasonably secure (especially since sales, even of the mild, are rising slightly) since the cask beers are, until the end of fermentation, identical to the 'Bobby Ales' keg products which are the mainstay of the business.

There are seventeen pubs, widely spread throughout the island, though five are in St Peter Port itself and three in St Sampsons. Seven have traditional beer, notably the Last Post in St Andrews, a busy and comfortable pub which is a rare outlet for the bitter, and the Mariners Inn, a traditional, basic bar near the quay in St Sampsons. In St Peter Port the Fermain Tavern and Jamaica Inn have real draught beer.

Jersey

ANN STREET BREWERY (Mary Ann)
The Ann Street Brewery Co Ltd, 57 Ann Street, St Helier, Jersey.
Pubs: 48. Very widespread free trade in Jersey. No traditional draught beers.

Although it is the largest of the Channel Islands breweries, producing more beer than the other three put together, and is owned largely by local shareholders (including the States of Jersey, which has a substantial holding), the Ann Street Brewery, founded in 1890, produces no traditional draught beers. It has two keg bitters, sold under the 'Mary Ann' trade name; Extra Special Bitter is reputed to be the strongest keg bitter in Britain. No mild has been produced since the 1950s – a sharp contrast to neighbouring Guernsey.

RANDALL'S (JERSEY)
Randalls Vautier Ltd, Clare Street, St Helier, Jersey.
Pubs: 27. Substantial free trade on the island. No traditional draught beers.

Randall's, which prides itself on being Jersey's oldest brewery, was founded by two brothers in 1823 and until just before the

First World War operated two breweries in St Helier. Production was then concentrated on the Clare Street site, and a completely new brewhouse was opened in 1937 – though within three years it was out of action again and remained so for the five years of the German Occupation. The occupying forces used the cellars as a wine and spirits store.

In 1963 the Randall family sold their controlling interest in the brewery to the Honourable Edward Greenall, of the Warrington brewing dynasty, and he remains the owner of the business. In 1967 he and Paul Clubb, the head brewer, developed Grunhalle lager, Edward Greenall having apparently had to buy a Bavarian brewery in order to acquire the recipe. Grunhalle is still brewed 'in the Bavarian manner' at Clare Street and is now also produced at a number of mainland breweries.

The bulk beers produced at Clare Street are two keg bitters, Local Bitter – another reminder of Greenall Whitley – and Top Island Bitter, two lagers and Whitbread Trophy, which is brewed under licence. Traditional draught beers were discontinued in the 1960s, and although in 1978 a new draught beer, Randall's Real Ale, became available, sometimes served direct from oak casks, it was withdrawn only four years later. Randall's Real Ale was the victim of falling demand, according to the brewery, or of the lack of any decent marketing effort, in the view of local real-ale enthusiasts. Whatever the cause the abrupt withdrawal of the beer (in any case closely related to Top Island Bitter) robbed Jersey of its only alternative real beer to Draught Bass.

Isle of Man

CASTLETOWN
Castletown Brewery Ltd, Victoria Road, Castletown, Isle of Man. Pubs: 34. Beers: Mild (1036), Bitter (1036).

Castletown Brewery is an excellent example of a company which is virtually unknown yet has to its name one of the finest draught beers in Britain, the quite magnificent draught bitter. There is one other oddity, too, for at one time it was owned wholly by another brewery company, and even now one of the Big Six has a thirty per cent shareholding.

But the beer comes first at Castletown, not least because the brewery (together with Okell's, its larger competitor) is subject to the Isle of Man's Pure Beer Act, dating from 1874. The Act states that, 'No ingredients other than malt, hops and sugar are used in brewing. A brewer is liable to a penalty of £300 and the confiscation of his entire plant and stock if this Act is contravened.'

Fortunately Castletown Brewery is in no danger of paying such a drastic penalty; its bitter (1036) not only conforms to the Act but is brewed with such skill that it is a really outstanding beer: a delicately balanced brew, with a dry, hoppy and above all light taste, it is worth the Isle of Man Steam Packet Company's fare from Liverpool all on its own. The bitter is available in traditional form in all but one of the tied houses, which are widely spread throughout the island; particularly recommended are the Bridge on North Quay, Douglas, which is a really excellent pub bursting with character, and the Union Inn in Castletown itself. Much of the beer is racked into wooden casks, and the bitter is dry hopped in the cask.

Castletown's Mild Ale (1036) is another matter. It accounts for a mere one per cent of production and is simply the bitter with caramel added to darken the colour. The result is a 'mild' without the true taste characteristics of the beer style, and frankly it is disappointing. It is, in any case, available in draught form in only a handful of the pubs. The mild also becomes Nut Brown Ale in bottle; the bitter doubles as Red Label Pale Ale; and the only 'different' brew is the winter-only Barley Wine, a malty and powerful (1074) bottled drink.

The history of the brewery is surprisingly complicated. There has been a brewery on the site, which has its own wells close to the river, for some hundreds of years – perhaps as far back as the fourteenth century. The oldest of the present buildings dates back to 1780, though the grey, battlemented tower which forms part of the brewhouse is much more recent. It was run by a succession of owners before the Castletown Brewery (1920) Ltd was incorporated in that year; at that time it was a private company controlled by the Cain family. In 1946–7, however, it was taken over by Hope & Anchor Breweries of Sheffield, famous for its Jubilee stout.

This could have been the end of the Castletown story, with the closure of the brewery a seemingly inevitable result of the takeover, but the reality was very different, for in 1948 the present company was formed to acquire not only the Castletown business but also Clinch's Brewery in Douglas. The Hope & Anchor shareholding was diluted in the process; this has now passed to Bass, who hold a thirty per cent stake but appear to leave Castletown to their own devices. Nevertheless, Bass' Carling lager is supplied to Castletown pubs, and occasionally Castletown brew Jubilee Stout under licence.

Some of this Jubilee Stout is intended for mainland consumption, a reminder that in the first half of the twentieth century Castletown's Ales of Man were quite widely available in Liverpool free houses. Sadly the Castletown ale and, to a large extent, the free houses are no longer part of the Merseyside scene, and a trip to the island is necessary to sample the excellent ale.

OKELL'S
Okell & Son Ltd, Falcon Brewery, Douglas, Isle of Man.
Pubs: 71. Free trade throughout the island. Beers: Mild (1035), Bitter (1036).

Okell's, the Isle of Man's largest brewery company, was founded by William Okell, a Douglas surgeon, at the Mount Falcon Brewery in 1850. Twenty-three years later he bought land half a mile away and built a new brewery for his son, William Henry Okell, and at around the same time he bought and closed down one of his local competitors, the Castle Hill Brewery. In 1877, in an extraordinary move, William Okell built the Lion Brewery in Waterloo, near Liverpool, in order to market his Light India Ale in England; by 1896 the Okells had abandoned this venture, and the brewery was known as Thoroughgood's, eventually being taken over by Threlfall's in 1925.

Back on the Isle of Man William Okell and his son went into partnership in 1890, closing the Mount Falcon Brewery and concentrating on the newer site, with its imposing and ornate Victorian brewery, which has remained virtually unaltered since 1875. The Okell family remained in control until 1972, when Heron & Brearley, a firm of wine and spirits merchants, bought the company. Heron & Brearley had previously bought pubs jointly with Okell's, and they had also acquired the pubs of Boddingtons' (Isle of Man) when the Manchester brewers disposed of their interests in the island.

Okell's draught beers – which, in contrast to Castletown's, are two entirely separate brews – are produced in strict accordance

A striking barrel label from Okell's Falcon Brewery

with the Isle of Man's Pure Beer Act (see p. 235) and comprise a very acceptable mild (1035.2), creamy and pleasant and accounting for some fifteen per cent of production, and a bitter (1035.9) which is very well hopped but at the same time has a smoother flavour than Castletown's noticeably acerbic taste. There is also a range of bottled beers, and certainly some of Okell's most memorable brews have been for bottling – notably the Coronation Ale in 1953 and Millennium Ale, an 1100 ale to celebrate the millennium of Tynwald, the island's parliament.

There are seventy-one tied houses, all except two of which sell traditional draught bitter, with a good proportion also offering draught mild. Many of the pubs are traditional locals with a warm welcome, although increasingly a programme of modernization is taking effect. In Douglas the British Hotel and the magnificent Edwardian Woodbourne Hotel are recommended; in Laxey the Mines Tavern is an excellent find; and the Creek Inn is a fine modernized pub in Peel. The free trade is also quite likely to offer Okell's beers, although these can be in keg form, especially in the many clubs and hotels which sell comparatively little beer.

Isle of Wight

BURT'S
Burt & Co, The Brewery, Ventnor, Isle of Wight.
Pubs: 11. Much more freely available in the free trade on the island. Beers: BMA Mild (1030), VPA Best Bitter (1040).

Beer has been produced on the site of Burt's brewery in Ventnor, nestling below the St Boniface Downs with its plentiful supply of chalky spring water, since at least 1840. Charles Richard Cundell was the owner until 1844, and his predecessor appears to have been Benjamin Mew, from one of the families which ran the island's other long-surviving brewery, Mew, Langton of Newport (taken over by Strong's in the 1960s). Cundell sold the Ventnor brewery to James Corbould in 1844, and Corbould also acquired three pubs which Burt's still own: the Central Hotel (then the Freemasons Commercial Inn) and Mill Bay in Ventnor, and the Stag Inn in Lake.

Corbould may not have been actively involved in brewing, since in 1850 John Burt and a Mr Keatley were described as tenants. By 1866 Frederick Corbould and John Burt had entered into a partnership, but this lasted only two more years, and finally in 1881 Burt became the sole owner. The Burt family's reign in absolute control was brief, however, because the brewery was acquired in 1906 by Albert Phillips. At the time of the take-over there were seven tied houses.

The brewery was inherited by William Arthur Phillips in 1913, and during the next thirty years a further five pubs were bought on a piecemeal basis, whilst one (the Wellington in Ventnor) was closed. In 1943 William Phillips was killed when the brewery was

A Burt's Strong Brown Ale label

destroyed by German bombs; yet within a year production had restarted in the ruins of the brewery. The completely new brewhouse which was required was completed in 1953.

Two beers are produced at the brewery, using only spring water, malt, hops and sugar, with some flaked maize the only additive. The spring water is supplied by the Isle of Wight water authorities for 2½p per annum, as a result of a remarkably far-sighted deal which James Corbould negotiated with the Ventnor Waterworks in 1850. The agreement allowed Burt's as much water as they needed for an annual rental of just 6d, and it has a lifespan of a thousand years. Attempts by the watermen to change the deal (Burt's use over a million gallons a year) have been successfully resisted. No wonder Burt's beer prices are the lowest in southern England!

The beers are VPA – Ventnor Pale Ale (1040), a malty and full-flavoured bitter without a particularly strong hop flavour, possibly because of the exclusive use of mellower Fuggles hops in the copper, and BMA – Best Mild Ale (1030), a fairly tasty but not especially memorable dark mild. Traditionally dispensed beers are in seven of the eleven tied houses, although until 1980 only two sold real ale; mild is rarely found in traditional draught form.

VPA accounts for by far the greater proportion of output, and it is almost exclusively this beer which is in demand from the island's free houses.

Bibliography

Students of the contemporary brewing industry are particularly well served by *What's Brewing*, the monthly newspaper of the Campaign for Real Ale, and by the *Good Beer Guide*. CAMRA also produces excellent polemical pamphlets such as *Whose Pint is it Anyway?* Of the books which have appeared over the last decade or so, those which I would recommend (though not from an unbiased standpoint!) are Frank Baillie's pioneering *Beer Drinker's Companion* (David & Charles, 1973), Chris Hutt's provocative *The Death of the English Pub* (Arrow, 1973) and my own, more factual, *Penguin Guide to Real Draught Beer* (Penguin, 1979). An invaluable source of information on brewery history is *Where Have All the Breweries Gone?* by Norman Barber (Neil Richardson/CAMRA, 1981). The best historical survey is H. S. Corran's *A History of Brewing* (David & Charles, 1975), though like many of the books on the beer world it has little to say about individual companies.

At the individual company level, several very good books have been written over the past decade, substantially raising the standard of brewery company histories from purely anecdotal to well-researched and weighty. Best among them, perhaps, are Richard Wilson's *Greene King: a Family and Business History* (Bodley Head, 1983); *Kimberley Ale: Hardys & Hansons 1832–1982* by George Bruce (Melland, 1982); *Hall & Woodhouse 1777–1977* by Hurford Janes (Melland, 1977); and Christopher Grayling's *Manchester Ales and Porter* (Joseph Holt, 1985). I also enjoyed Francis Shepard's *Brakspear's Brewery, Henley on Thames 1779–1979* (Brakspear's, 1979) and *The Mansfield Brew* by Philip Bristow (Navigator Publishing, 1976). Several other breweries, such as Boddingtons', Greenall Whitley, Higson's, Oldham Brewery, Simpkiss's, Wadworth's and Young's, have also produced good books or pamphlets, often to celebrate a particular milestone in the company's history.

Index

Adnams, 19, 20, 40, 141, 142–4
Air pressure, 52, 223, 227
AK, 69–70, 165
'Ale Trail', 68
Allied Breweries, 31, 34, 57, 108, 118, 157, 167, 174–5, 176, 190, 204, 216, 220, 222, 224, 225
All Nations, Madeley, 29, 118, 119–20
Ancient Druids, Cambridge, 76
Ann Street Brewery, 228, 233
Ansell's, 31, 92, 113, 118, 125
Archer's, 81
Arkell, Claude, 81, 86–8
Arkell's, 54, 55, 80, 81–3

Badger Beers, *see* Hall & Woodhouse
Baillie, Frank, 241
Banks's, 52, 53, 108, 118, 119, 120–3, 165
Barber, Norman, 241
Barclay Perkins, 25
Barley wine, 54, 131, 134, 201
Bass, 25, 26, 31, 36, 44, 48, 57, 66, 80, 108, 134, 150, 171, 173, 186, 188, 206, 220, 226, 228, 234, 236
Bateman's, 40, 49, 53, 54, 141, 144–5, 223
Batham's, 20, 22, 40, 54, 118, 123–5, 127
Beard's, 65
'Beer at Home', 125, 126
Beer from the wood, 19, 72, 106, 184, 194, 200, 209, 210
Belhaven Brewery, 38, 39, 42, 108, 112, 220, 221–3
Bell Brewhouse, Shoreditch, 24
Big Six, 30–2, 36, 57, 80, 108, 131, 141, 173, 204, 220, 225, 234
Birmingham, 29, 118

Bitter, 53–4
Blackawton Brewery, 81
Black Country, 53, 118, 120, 123, 127, 134–6, 137
Blanket pressure, 52–3, 82, 90
Blue Anchor, Helston, 29, 80, 81, 83–4
Bluebell, Hockley Heath, 30
Boddingtons', 31, 34, 36, 47, 120, 173, 174–7, 185, 187, 188, 197–9, 228, 236, 241
Border Breweries, 34, 41, 97, 108, 131, 133, 179
Bottled beers, 55–6, 60, 73, 158–9, 184, 208
'Boy's bitter', 54, 101, 104
Bradford, William, 39, 65, 130
Brain's, 27, 38, 54, 108, 109–11
Brakspear's, 27–8, 36, 42, 54, 55, 58–60, 143, 241
Bramah, Joseph, 51
'breweriana', 55
Brewex, 78, 167
Brewing sugar, 44
Brewing water (liquor), 42
Bristow, Philip, 241
Britannia Inn, Loughborough, 30
Brown, Matthew, 33, 34, 36, 160, 162, 173, 174, 177–8, 192, 202, 204, 213, 214–15
Broughton Brewery, 221
Bruce, George, 241
Buckley, Revd James, 20, 111
Buckley's, 36, 43, 44, 108, 111–13, 116, 223
Bull & Bladder, Delph, 40, 123, 125
Burt's, 228, 238–40
Burton-on-Trent, 17, 19, 25, 26, 40, 42, 47, 57, 118, 147
Burton Union system, 19, 47–8, 49, 131, 134

Burtonwood Brewery, 20, 22, 36, 50, 108, 130, 133, 173, 174, 179–81

Cameron's, 46, 148, 170, 204, 205–7
Campbell, Hope & King, 41, 220–1
CAMRA, 32, 60, 111, 135, 145, 148, 241
Canned beer, 116
Carriage casks, 48–9, 112
Cask breather, 52, 90, 98
Cask sizes, 50
Castletown Brewery, 36, 41, 228, 234–6, 237
Charrington, 25, 31, 57, 80, 162, 220, 226
Chester Northgate Brewery, 168, 182
Chester's, 31, 173, 174
Christmas & Co, 149, 152
Circus Brewery, 139–40
Clark's of Wakefield, 205
Clinch's Brewery, 47, 236
Clore, Sir Charles, 31
Clubs' breweries, 108, 113–15, 205
Common brewers, 22, 24–6, 28–9
Conditioning, 49
Cook's, 73
Cooper's shops, 19, 106, 200, 210
Coppers, 43–4, 117
Corran, H. S., 241
Country pubs, closure of, 60, 86, 98, 102, 143, 205, 206
Courage, 25, 31, 32, 40, 56, 57, 80, 129, 204, 209, 223
Crown Brewery, 108, 113–15

Darley's, 19, 47, 204, 207–9, 215–17, 219
Davenport's, 34, 49, 118, 120, 125–7, 181, 183
Desnoes & Geddes, 76
Devenish, 36, 54, 80, 84–6
Donnington Brewery, 38, 39, 40, 42, 80, 81, 86–9
Druid's Head, Coseley, 30
Drybrough's, 220

Dry hopping, 50, 54, 61, 122, 138, 215
Dudgeon & Co, 221

Edinburgh, 40, 41, 220
Eldridge Pope, 17, 22, 33, 37, 50, 54, 56, 80, 81, 85, 89–90, 101
Electric pumps, 51–2, 122–3
Elgood's, 17, 19, 39, 46, 53, 141, 145–7
Ellerman Holdings, 34, 141, 170–1, 204, 205–6
Everard's, 33, 58, 112, 118, 141, 147–8, 167, 207

Falkland Islands, 148
Felinfoel Brewery, 36, 43, 108, 112, 115–17
Fermentation, 46–9
Finings, 51
Flower's, 150
Fremlin's, 41, 57, 74
Fuller's, 33, 36, 39, 50, 53, 54, 58, 60–2, 79, 109
Fun pubs, 169

Gale's, 36, 44, 46, 54, 56, 57, 62–4
Gibbs Mew, 80, 91–2, 162
Gipsie's Tent, Dudley, 30
Golden Cross, Cardiff, 110–11
Golden Hill Brewery, 80–1
Golden Lion, Southwick, 30, 41
Good Beer Guide, 120, 134, 241
Grand Metropolitan, 32, 34, 126, 191
Grayling, Christopher, 241
Gray's, 73
Great British Beer Festival, 60, 81, 201
Greenall Whitley, 118, 125, 126–7, 136, 141, 160, 168–70, 173, 174, 181–3, 234, 241
Greene King, 20, 22, 54, 58, 70, 73, 109, 141, 148–52, 163, 165, 171, 241
Groves, John, 80, 85
Groves & Whitnall, 168, 182
Guernsey Brewery, 40, 49, 229–31
Guinness, 30, 31, 56, 91, 150, 171

INDEX

Hall & Woodhouse, 80, 81, 93–6, 241
Hall's, 80, 99, 167
Handpumps, 51
Hanson's, Julia, 30, 52, 53, 118, 119, 120–3
Hardy's & Hanson's, 18, 20, 36, 39, 47, 141, 152–5, 241
Hartley's, 173, 183–5, 199, 200
Harvest ales, 130, 162
Harvey's, 17, 39, 47, 50, 54, 57, 64–6
Henley Brewery, *see* Brakspear's
Heritage Brewery, 147
Highgate Brewery, 118–19
Higson's, 34, 39, 173, 174, 176, 180, 185–7, 197, 241
Hogsheads, 50, 64, 190
Holden's, 54, 118, 119, 122, 127–9
Holt's, 27, 50, 54, 55, 120, 173, 187–90, 241
Home Brewery, 141, 156–8
Home-brew kits, 82, 162
Home-brew pubs, 22, 29–30, 83–4, 105, 116, 118, 119–20, 123, 129, 134–6, 137, 138–9, 157, 218
Hook Norton Brewery, 19, 34, 36, 39, 40, 44, 46, 65, 118, 129–31
Hope & Anchor Breweries, 36, 204, 220, 228, 235–6
Hops, 23, 38, 41, 42, 43–4, 50, 98
Horse-drawn drays, 19, 51, 68, 97, 106, 143, 200–1, 217, 229
Hoskin's, 141, 142, 158–60
Hull Brewery, 34, 161, 171, 199
Hunt Edmunds, 31, 129
Hutt, Christopher, 241
Hyde's, 52, 53, 173, 190–1

Imperial Group, 32
Ind Coope, 25, 31, 61, 70, 118, 159, 162, 228
Inn Leisure, 85

Janes, Hurford, 241
Jennings, 31, 34, 36, 40, 173, 191–3

Keg beer, 43
King & Barnes, 32, 50–1, 53, 54, 57, 66–8

Lager, 53, 89, 100, 177, 185, 187
Lancashire Clubs' Federation Brewery, 91
Leeds, 29
Lees, 19, 20, 50, 108, 173, 188, 193–5
Liquor, 42
Lorimer & Clark, 19, 44, 215–17, 220, 221, 224–5

Maclay's, 36, 220, 221, 225–7
Magee Marshall, 42, 182
Maidstone, 40, 57, 74
Malt, 41
Maltings, 38, 42, 72, 101, 106, 111, 121, 129, 130, 154, 223
Mann, Crossman & Paulin, 25, 132
Mansfield Brewery, 34, 44, 47, 141, 160–1, 169, 241
Marston's, 19, 25, 33, 36, 39, 41, 42, 47–8, 54, 97, 108, 118, 131–4, 179
Mashing, 43
Matthews & Co, 80, 96
McEwan's, 220
McMullen's, 58, 69–71, 129
Melbourn's, 35, 167, 210
Metropolitan Clubs' Brewery, 40, 114
Mew Langton, 129, 228, 238
Mild, 53, 82, 86, 118–19, 131, 134, 143, 146, 151, 174
Miss World, 222–3
Mitchell's, 173, 195–7, 202
Mitchell's & Butler's, 118, 125, 129
Monastic brewers, 17, 39, 74, 97, 99, 223
Monopolies & Mergers Commission, 177, 206
Monopolies, local, 141, 178
Morland's, 33, 36, 54, 80, 81, 96–8, 229
Morrell's, 17, 39, 40, 54, 80, 81, 98–100

Mount Charlotte Investments, 31, 178, 191

National Brewery Museum, 147
Netherton Ales, 135, 136
North Country Breweries, 160, 161, 167
Northern Clubs' Federation Brewery, 114, 205
Northern Foods, 34, 161, 169, 171, 199
Northern United Breweries, 220, 225, 226

Oast-houses, 38
Okell's, 17, 39, 41, 228, 235, 236–8
Old ale, 54, 66, 68, 79, 127, 137, 143
Old Speckled Hen, 98
Old Swan, Netherton, 19, 29, 46, 118, 127, 129, 134–6
Oldham Brewery, 34, 173, 174, 176, 197–9, 241
Open coolers, 44–6, 64, 131
Original gravity, 54–5

Paine's, 46, 92, 141, 162–3
Palmer's, 37, 40, 80, 100–2
Paraflow refrigerators, 46
Pardoe, Doris, 19, 20–1, 134, 136, 140
Penguin Ale, 148
Piggins, 145
Porter, 22, 24, 42, 212
Priming sugars, 50–1
Prize Old Ale, 56, 64
Publican brewers, 24, 27, 28–30
  see also Home-brew pubs
Pure Beer Act, 41, 235, 237
Puzzle Hall Brewery, 218

Queen's College brewhouse, Oxford, 99

Randall's (Guernsey), 228, 231–3
Randall's (Jersey), 229, 233–4
Rayment's, 141, 149, 151, 163–6
Redruth Brewery Co., 85, 104
Reinheitsgebot (Purity Law), 41

Ridley's, 53, 58, 71–3
Robinson's, 53, 54, 108, 173, 183–4, 188, 199–201
Ruddle's, 33, 54, 57, 109, 141, 147, 166–8, 211

St Austell Brewery, 20, 44, 50, 54, 80, 81, 102–4
Scotch ale, 216, 217, 224
Scottish & Newcastle, 31, 32, 34, 171, 173, 177, 204, 206, 220
Selby Brewery, 204–5
Shepard, Francis, 241
Shepherd Neame, 17, 33, 54, 58, 73–5, 94
Shipstone's, 34, 141, 160, 168–70, 181, 183
Shrewsbury & Wem Brewery Co., *see* Wem Brewery
Simond's, 59, 96, 150
Simpkiss, 34, 118, 129, 136–7, 139, 183, 241
Simpson's, 149, 151, 165
Smith's, John, 204, 209
Smith's, Samuel, 19, 36, 43, 47, 50, 167, 192, 204, 209–11, 215
South Wales & Monmouthshire United Clubs' Brewery Co., 114
Spingo ale, 84
Stag Brewery, Pimlico, 25
State Management Scheme, 173–4, 193, 213
Steam engines, 77, 131
Steward & Patteson, 28
Stock ale, 54, 75
Street, A. G., 20
Strength of beer, *see* Original gravity
Strong ale, 54
Strong's, 57, 228, 238
Style & Winch, 40
Supermarket beers, 76, 141, 166, 167, 177

Taylor, Eddie, 220
Taylor's, Timothy, 34, 54, 204, 211–13
Tennent's, 192, 220

INDEX

Tetley, 31, 180, 193, 204
Tetley Walker, 173, 190
Theakston's, 33, 34, 54, 174, 177, 204, 213–15
Thomas Hardy Ale, 56, 89, 90
Three Crowns, Devizes, 105
Three Tuns, Bishops Castle, 29, 118, 138–9
Threlfall's, 31, 236
Thwaites, 49, 53, 58, 173, 197, 201–3, 230
Tied houses, 22, 26–7
Tolly Cobbold, 20, 34, 141, 149, 162, 170–2, 206
Top pressure, 53, 86, 90, 98, 145, 149
Traquair House, 221
Truman's, 26, 57, 73, 149, 211

Usher's (Edinburgh), 216, 224
Usher's (Trowbridge), 59, 80, 166

Vaux, 19, 20, 22, 51, 58, 173, 188, 204, 206, 207–8, 215–17, 220, 224, 225
Virani, Nazmu, 112, 223

Wadworth's, 17, 19, 20, 33, 39, 44, 46, 50, 54, 80, 112, 241
Ward's, 52, 204, 208, 215, 216, 217–19
Waterwheels, 40, 88, 100, 101
Watney's, 25, 28, 31, 32, 57, 62, 66, 80, 141, 167, 171, 173, 191, 220
Webster's, 204
Wells, Charles, 20, 58, 75–7
Wells Fargo, 76
Wells & Winch, 150
Welsh Brewers, 108
Wem Brewery, 108, 118, 137, 139–40, 181–3
West Country Breweries, 31, 80
Wethered's, 57, 59
Whitbread's, 25, 26, 32, 33, 34, 36, 41, 57, 60, 62, 74, 80, 97, 108, 112, 113, 131, 133, 147, 148, 150, 166, 167, 171, 173, 176, 179, 187, 188, 202, 205, 220, 228, 234
Whitbread Investment Company, 36, 133, 177
Wilson, Richard, 149, 241
Wilson's Brewery, 122, 173
Wolverhampton & Dudley, 38, 53, 118, 120–3, 125, 126–7, 133;
  see also Banks's and Hanson's
Wooden casks, *see* Beer from the wood
Workington Brewery, 34, 173, 177, 178, 191
Wort, 43–6, 48
Worthington, 25, 26, 30, 56

Yates & Jackson, 34, 173, 195, 197, 201, 202
Yeast, 42–3, 46, 48, 112, 122, 143
Yorkshire Clubs' Brewery, 35
Yorkshire squares, 19, 47, 49, 207–8, 209–10
Young's, 19, 32, 33, 49, 51, 54, 58, 77–9, 241
Younger's, 220

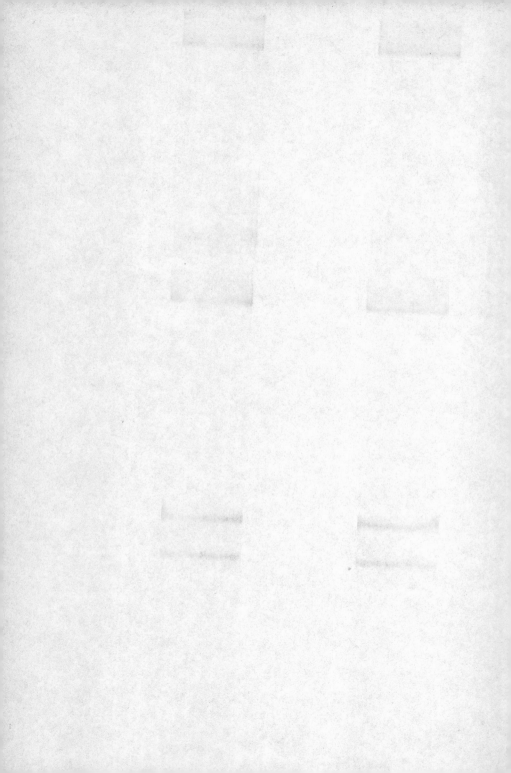